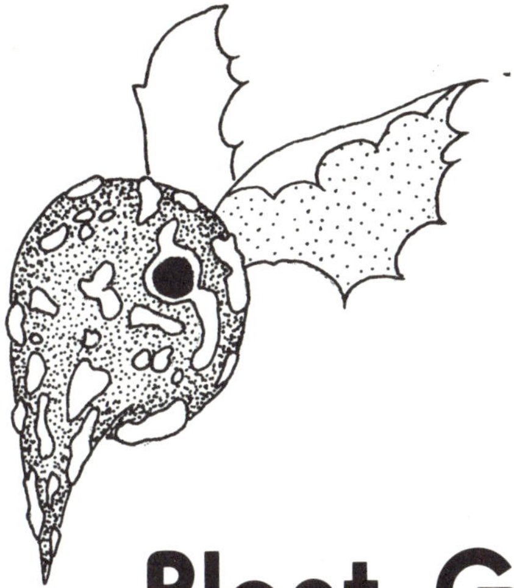

Plant Galls of the California Region

by

RONALD A. RUSSO

Photographs and Illustrations by the Author

THE BOXWOOD PRESS

Distributed by:

The Boxwood Press
183 Ocean View Blvd.
Pacific Grove, CA 93950
408—375-9110

Cover design
by
Pam Olhausen

Library of Congress Cataloging in Publication Data Russo, Ronald A.

Plant galls of the California region.

Bibliography: p.

Includes index.

1. Galls (Botany)—California—Identification. 2. Gall-flies—Identification.
3. Gall-gnats—Identification. 4. Insects—California—Identification.
Title.

SB767.R87 582'.02'33 79-20506

ISBN: 0-910286-71-X

Printed in U.S.A.

DEDICATED
to
my three sons,
Ron, Ken, and Brian.

FOREWORD

THE APPEARANCE of this book coincides appropriately with a period in the American lifestyle which brings indoor people out of doors and into forests, parks, and other recreational lands. Never before have more people had a greater interest in their surroundings and in the plants, animals, and other elements of which they are comprised. Unlike generations of a century ago, it is not now economics or survival that encourages familiarity with one's surroundings; instead, it is an escape from an increasingly urban life, knowledge that open space is becoming less accessible, and a general belief that one ought to know more about nature. This book will help to clear up some of the mysteries of plant abnormality most persons have noticed at one time or another, and allow the outdoors to become a bit more enjoyable and interesting.

As subject matter for primary and secondary school science education, galls and gallmakers have been vastly underutilized. They are well distributed throughout California, easy to collect—and now with this book, easier to identify—and should become a more important part of school biology curricula. Further, galls and gallmakers involve interaction of uncommon intimacy between the plant and animal kingdoms.

Not since the time of Lewis Weld has anyone but Ron Russo had quite the courage to attempt a comprehensive, lay treatise on California's galls and gallmakers. The group of organisms responsible for galls lacks the popular appeal of, say, the butterflies and beetles, and the economic justification of the moths, aphids, and flies for their study. Surprisingly little is yet known about the exact mechanisms by which galls are initiated by their makers, and even less is documented to explain the variety of shapes, colors, and other attributes of the plant tissue proliferations we term galls. Despite the fidelity of specific gallmakers to the formation of a precise gall, the exact organismal form responsible for many galls, especially those with alternating generations, remains elusive. All signs indicate that this state of affairs is not likely to change in the immediate future.

C.S. Koehler

University of California/Berkeley

PREFACE

MY ODYSSEY with plant galls began on a dry hillside near Martinez, California, in 1970, when I discovered some incredibly beautiful purple and red spiny wasp galls on blue oaks. I was so moved by the diversity of species found on that field trip, that I felt compelled to explore further and learn as much as possible. Motivated by simple curiosity, I found myself traveling throughout California collecting specimens, field notes, and photographs. Immediately apparent to me was the fact that such extraordinarily beautiful creations were so abundant, yet, so collectively ignored.

I vividly recall lying awake all night during a rainstorm in Sequoia National Forest wondering what I was going to do with my enthusiasm for the subject and all of what I was discovering. Out of the need to share with others a part of the magic and excitement of this area of our natural resources, the idea of this book was born. It is intended to open new doors for some and create greater understanding among others who might choose to pick this book up. It is presented with the hope that those who use it will find the same exhilaration and wonder that I have experienced.

This work represents the first attempt in the California area to portray the biology and taxonomy of gall organisms on a broad scale. It is by no means complete or definitive. This book is a beginning, an introduction to a group of organisms of which many species remain unidentified. With continuing interest and time, I hope that the scope of this book can be broadened in future revisions. Although I have tried to include as much biological and ecological data as possible, there still exist great voids that only continued research on the part of many people can fill. As it stands, the applicability of this work extends well beyond the political boundaries of California. Many of the host plants mentioned here occur elsewhere. In addition, many of the gall organisms are not restricted to a single species, but occur on related species in the western United States.

Though scientific names are indispensable handles for botanical researchers, forest pathologists, and most naturalists, they are often difficult to learn and easily forgotten by teachers of the young. Whenever I could I have attached common names, many not previously mentioned in the literature. I hope that this will not be considered a breach of scientific etiquette.

Galls are one of the many elements of our natural world that have never enjoyed the popularity of birds, wildflowers, or mushrooms. Yet there is great beauty, in the form and color of galls, that should appeal to naturalists and also to artists and photographers of nature. However, the esthetic appeal of galls must not overshadow the tiny creatures inside, whose lives are important to the scientists who specialize in the study of galls and also to ecologists who try to understand all the elements of ecosystems.

ACKNOWLEDGEMENTS

I am indebted to L.E. Caltagirone and C.S. Koehler of the University of California, Berkeley for their assistance and suggestions in reviewing the manuscript. I also appreciate the suggestions and critical review of appropriate sections by W.H. Lange (aphids), U.C. Davis; C.E. Yarwood (fungi), U.C. Berkeley; H.H. Kiefer (mites), Raymond Gagne (midges), U.S. National Museum; Charles Dailey (cynipid wasps), Sierra College; and Wayne Roderick (botany), East Bay Regional Park district. I would also like to extend a note of thanks to Susan Reed for her review of the introductory text and suggestions for organization. A special note of thanks goes to Alan Kaplan who reviewed and commented on the manuscript, to Howell Daly who encouraged me in this project, to Pam Olhausen for her illustration of the woodpecker, and to Al Seneres and M. L. Wilcox for contributing specimens to my collection.

The final draft was prepared on the basis of the suggestions and criticisms offered by the reviewers. But, of course, any inaccuracies or inadequacies in this book are the responsibility of the author and not the reviewers.

CONTENTS

1

INTRODUCTION

IF you have ever hiked or camped beneath oak trees or if you have oaks in your garden, you probably have noticed peculiar swellings or deformities on the leaves or branches. These swellings are plant galls. A gall is the result of an intriguing reaction by a plant to the invasion of organisms such as certain fungi, mites, aphids, flies, and wasps. .

The world of plant galls is a lilliputian realm in which an organism scarcely the size of a period on this page can induce a plant to produce a swelling of such flamboyant design and color that it would challenge the wildest of architectural dreams. In their own right, gall organisms are nature's miniature but master architects. A part of the "magic" of this little-known world is that plant galls and gall organisms are all around us.

Along the foothills, in the great Central Valley, and the mountains of the Sierra, Southern Cascade, and Coast Ranges in California, for example, a rich variety of oak trees sports an equally rich and fascinating display of some of the most oddly shaped and brightly colored galls to be found anywhere. Interestingly enough, the most beautiful galls in the western United States can be found as easily in the backyard as in the wilderness. Galls are found on many native shrubs and trees, as well as on a few cultivated plants. It takes but a dash of curiosity to explore, discover, and realize the fascination of these strange creations.

Plant galls are marked by the vast diversity of sizes, shapes, and colors. They range in size from that of a pinhead to a volleyball. In color galls present a broad spectrum, from subtle hues of beige and green to striking shades of pink, purple, and orange. Some galls are round and smooth, resembling marbles or ping-pong balls. Other galls are spiny or densely hairy and may resemble sea urchins or caterpillars. Some are so bright that one cannot help but notice them. Others, because of their size or color, defy even the most careful observer. Few people who notice galls understand the intricate nature of the relationship between the host plant and the cause of the swelling.

Even though certain galls are well known to industrial and agricultural interests as sources of tanning agents, ink, and supplementary livestock feed, to most people who find them galls represent another of the many unexplained natural curiosities. The purpose of this book is to acquaint the reader with the causes, effects, interrelationships, and identity of some of the galls found in California. This is not a book on insects nor is it a book on plants. Instead, this book endeavors to interpret one of the strangest relationships between plants and plants, and plants and animals.

The galls that result from these highly refined relationships can tease the imagination and capture the fancy of anyone who dares to investigate but a single, prize specimen. In studying galls, one may catch the same kind of fever that has focused the attention of millions of people on the observation and photography of rocks, birds, and wildflowers. Like these interests, the investigation of galls can be one of a series of lifelong adventures with the natural world.

THE SCIENCE OF GALL STUDY

CECIDOLOGY is the study of plant galls and the organisms that induce them. This ecological science brings together botany, entomology, and parasitology in close association. Plants, insects, and organisms that feed on both are important elements in the world of galls. Because the gall-inducing organism is often extremely selective in choosing the species of plant on which to deposit its eggs or to engage in feeding, one can occasionally identify the kind of plant by the type of gall, or the gall by the type of plant. The scientific names given to the galls are actually the names of the gall organisms responsible for the swellings. Since many gall organisms are scarcely visible, it is easier to identify the gall inducer by the gall than by the organism itself.

WHAT ARE GALLS?

GALLS are tumorlike growths composed of either plant cells that have undergone multiplication in abnormally high numbers or of greater size than normal, or plant organs such as buds and fruit whose growth and development have been altered. The cells of some galled plants differentiate into types not normally found in the ungalled plants. These types of galls are commonly associated with cynipid wasps. The galls often have bizarre shapes. Other galls are composed of cells that hardly differentiate and may form into simple swellings. Nematine sawflies are associated with these types of galls.

The formation of a gall is initiated by the chemical or mechanical stimulus of an invading plant or animal. In some cases the stimulus to gall formation may actually be a combination of both chemical and mechanical processes. The chemical stimulus leading to gall formation may originate when a gall insect such as a tenthredinid wasp stings a leaf and injects a fluid rich in nucleic acids and protein along with its eggs. Nucleic acids are important in conveying genetic information. Since the gall that results from the action of the wasp on the plant will be of a specific size, shape, and color characteristic to that species of wasp, it is thought by some scientists that the chemicals produced by the insect genetically programs the development of the plant tissue in forming a particular gall. The stimulus to gall formation may take place only after the egg hatches and the larva begins feeding on the tissues of the plant. The mechanical stimulus of chewing, as well as chemicals released by the larva, may activate or further stimulate the growth of the gall. In some cases, noticeable galls do not form until the mechanical irritation is provided. Unfortunately, not enough research in this area has been conducted to know the precise operating mechanism of most gall organisms.

The reaction of the plant to the activities of the invading organism is one of defense and constitutes an attempt to isolate the toxins or activities of the invader. Ironically, in doing so the plant provides food and some degree of shelter for its enemy.

Galled plant tissue normally swells or expands outwardly from the area of the egg or activity of the invading organism. This growth pattern usually leaves a chamber or cavity, which becomes the feeding territory of the organism. Sometimes, however, because of rapid cell division, growth occurs inwardly, crushing the organism.

WHERE GALLS FORM

VIRTUALLY every part of a plant may be subject to gall formation. Whereas galls form on the roots, new branches, buds, leaf petioles, leaves, flowers, and fruit of such plants as oak trees, other plants may support galls only on the leaves or branches.

More than 50% of the known galls occur on leaves rather than on other plant parts. Organisms such as insects seem to find a wider choice of conditions suitable for their survival on leaves than on other plant organs. High photosynthetic activity and rapid growth are among the chief factors that have led to the wealth of leaf galls.

At different times of the year a single tree may support galls on most of its organs. Coast live oak, *Quercus agrifolia,* is an example. Close examination of a coast live oak in the spring might reveal the wasp galls of *Callirhytis congregata* on the flowers or catkins, and *C. quercussuttoni* on the branches. The galls of *C. flora* may be found on the midrib veins or petioles of new leaves. Later, in the summer, one might find the galls of *Dryocosmus minusculus* on the tops of the leaves, and the galls of *D. dubiosus* on the veins of the undersides of the leaves. In early summer the greenish, fuzzy galls of *Callirhytis agrifoliae* form at the tips of the branches. Not every oak in a woodland would host all of these gall wasps during the course of the year. Some oaks of the same species might support only one or two wasp species. However, it would indeed be a rare occasion to find an oak that did not support a single gall organism.

COMMON TYPES OF GALLS

THE MOST COMMON types of galls found by the average observer usually occur on the leaves and branches. Leaf galls are divided into specific types based on their general structure. There are roll galls, fold galls, pouch galls, and mark galls. **Roll** and **fold galls** are somewhat similar in basic design. Some organisms cause the edge of the leaf to roll over, creating an enclosure. Other organisms cause the leaf to fold in the middle along the midrib

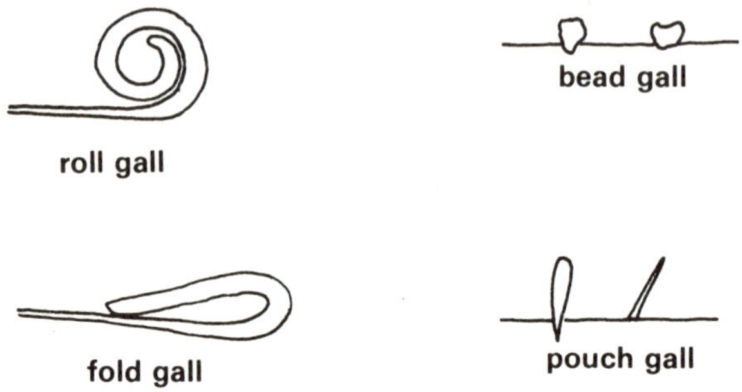

Fig. 6

vein or at the edge, but in a single folding layer. **Pouch galls** often appear as oblong, club-shaped growths with an opening or **ostiole** at the base. Other pouch galls are globular beads on leaf surfaces. **Mark galls** are characterized by the fact that the gall organism is initially enclosed within the gall. Mark galls may hang or protrude from the leaf surface and be easily detached, or they may form as integral, blisterlike swellings that are not easily detached without damaging the leaf surface. Another type of leaf gall, particularly on oaks, is referred to as an **erineum,** and appears as a hairlined depression on the leaf surface. In the case of erineum galls or **filzgalls,** as they are also known, the gall organism may or may not live inside the plant tissue.

In looking at the stems and branches of oak trees, one finds a much simpler division of gall types. **Stem** and **branch galls** are either detachable or nondetachable. They either protrude and are easily removed or are integral swellings impossible to remove without cutting the stem. The California gallfly, *Andricus californicus,* which is a wasp, induces a large, lightweight gall that protrudes from the branches of valley oaks, *Quercus lobata.* Although these galls may remain on the tree for years, they are easily removed. The galls of *Dryocosmus asymmetricus* are elongated, integral swellings that rupture at maturity, but last on the tree as long as the effected branches. Most of the stem and branch galls are relatively unattractive compared to the oddly shaped and brightly colored leaf galls.

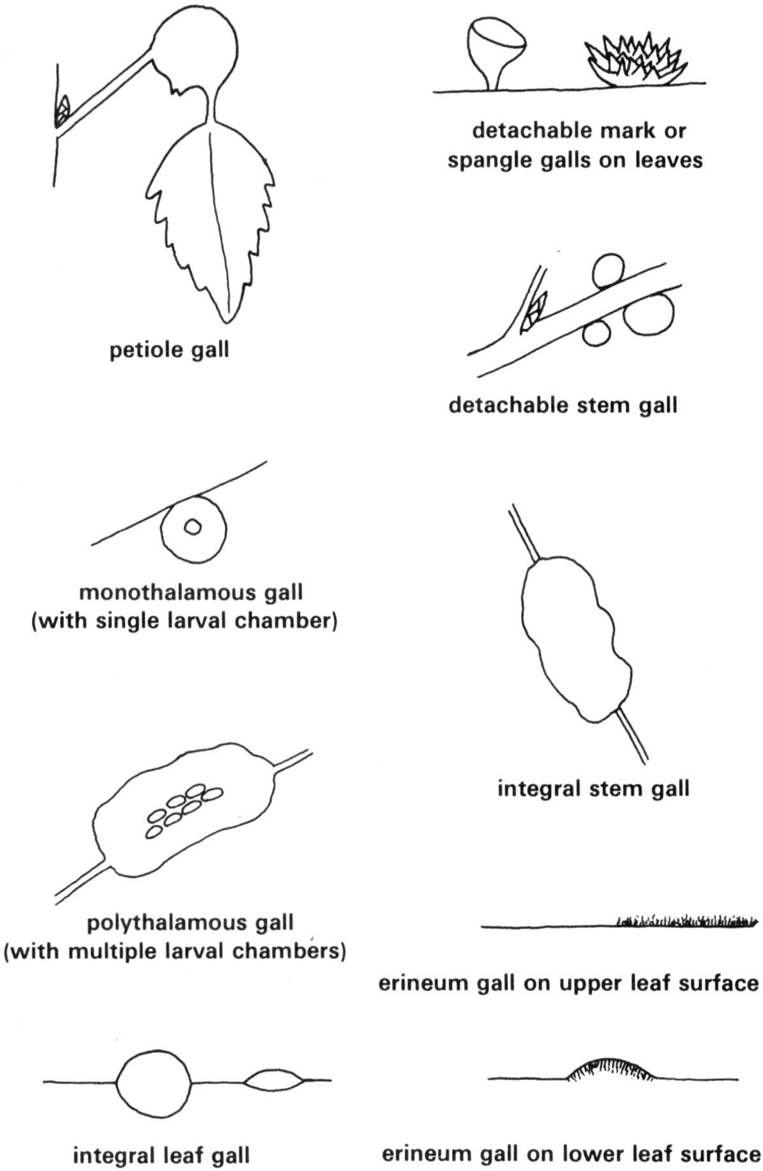
petiole gall
detachable mark or spangle galls on leaves
detachable stem gall
monothalamous gall (with single larval chamber)
integral stem gall
polythalamous gall (with multiple larval chambers)
erineum gall on upper leaf surface
integral leaf gall
erineum gall on lower leaf surface

Fig. 7

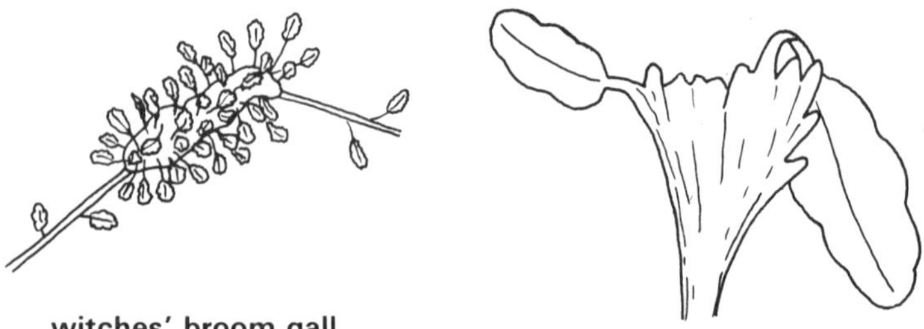

witches' broom gall

fasciated terminal bud gall

Fig. 8

In addition to the types of galls discussed there are other forms that deserve mention either because they are in fact galls or because they are confused with galls. **Witches' broom** galls are dense clusters of shoots and stems emanating from a common focal point. These galls can result from the attacks of fungi, viruses, bacteria, mites, and other organisms. Those witches' brooms that originate from mechanical injury not associated with an organism are not true galls. Organism-induced witches' brooms are common on oaks, baccharis, and pines. In many cases, the causative organism is not known.

Another gall-like structure is a **fasciation.** Fasciations are enlarged, flattened, broad sections of the terminal areas of branches and floral heads. In some cases mites and insects may be involved. However, in other cases it is suspected that genetic changes within the plant cause the fasciations. These fasciations occur on a wide variety of native and ornamental plants. I have found massive, fasciated basal shoots surrounding the entire circumference of large coast redwoods, *Sequoia sempervirens,* but the exact cause has yet to be determined. The common mullein, *Verbascum thapsus,* found growing along railroad tracks and gravelly roadbeds and river banks, sports some of the largest fasciated floral heads I have ever seen. They may reach 15-20 cm across, whereas the normal columnar floral spikes measure about 5-7 cm in diameter.

Burls, which are knobby outgrowths with unusual grain near the base of many trees, are often highly prized as curio items or as sources of wood for furnishings. Burls have been referred to as nothing more than trunk galls. This is open to debate. The burls associated with the bases of coast redwoods and bay laurel, *Umbellularia californica,* result from either dormant buds or adventitious buds and are not apparently true galls. Years ago at Big Basin State Park I found and measured a trunk gall several meters from the ground that spanned over 110 cm in diameter. Its rough, fissured appearance was much like that of canker galls found on aspens and cottonwoods. In a grove of redwoods on the University of California campus at Berkeley the branches of a single redwood are literally covered with knobby, gall-like growths. Although attempts were made to determine the cause, the nature of these growths has yet to be discovered. Along with other organisms previously mentioned, viruses, which are difficult to isolate, may play a role in some of the seemingly unexplainable growths found on redwoods and other trees.

SEASONAL APPEARANCE AND GROWTH RATE

THE APPEARANCE of plants and animals in nature is influenced by many factors, including changes in temperature, moisture, length of daylight hours, and, in the case of gall organisms, the availability of food and nesting habitats. The appearance of many plant galls coincides with the season of greatest plant growth.

In the spring and early summer, depending on altitude, vast amounts of food energy are directed towards the production or expansion of buds, new shoots, branches, and leaves. During this season many galls seemingly appear overnight. Actually, their growth requires much more time. Under the influence of the proper stimulus, energy is rerouted into the development of gall tissue. In the summer at low elevations, most plant growth slows down. At this time energy that would normally sustain already existing plant organs may be redirected to the development of a second flush of galls. The growth of the summer galls normally proceeds at a much slower rate than that of the spring galls.

An individual blue oak, *Quercus douglasii,* may produce both a spring and a summer crop of galls. Actually a gall produced in the spring may belong to the same species of wasp that emerges from a particular summer-generation gall. Even though the two generations of galls may differ greatly in size, shape, and location, they may be produced by different generations of the same insect. Many species have different, alternating generations that induce different galls on different plant parts at separate times of the year. This alternation of generations, known as **heterogeny,** results in different modes of reproduction. Often the insects hatching from a specific type of gall in the spring are males and females, and reproduce sexually. After the females of the sexual generation deposit their eggs, a second crop of galls induced by this species develops and matures between summer and fall. From these summer and fall galls emerge only females, which reproduce without the aid of males. This phenomenon, known as **parthenogenesis,** is common among many insects. As a result of the actions of this unisexual generation of females, galls are induced that will eventually issue the bisexual generation. For many gall organisms there are no males known for the particular species, and reproduction is parthenogenetic. This subject will be explored at greater length in examining the groups of gall-inducing organisms.

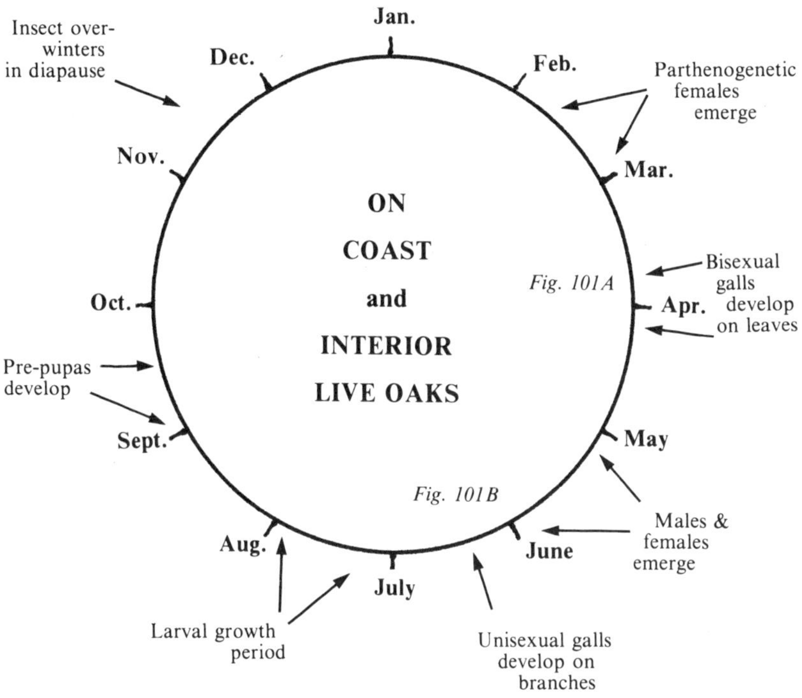

Fig. 11. General life cycle of the cynipid wasp *Callirhytis pomiformis,* in which two separate generations produce distinct galls at different times of the year. Unisexual galls can remain on trees after adults emerge. Variations occur among different localities and altitudes.

ENVIRONMENTAL FACTORS AFFECTING GALLS

WITHIN A SPECIES of tree or shrub, some individuals are relatively barren of galls while others support a large number. The reasons for this variation involve many physical factors, such as full exposure to sun and high temperatures, shade, humidity, and soil characteristics, that influence the physiology of the plant. A willow tree growing in a warm, dry, south-westerly exposed situation may not support the diversity and quantity of gall organisms as the same species growing in a cooler, damper, north-easterly exposure. The thickness of the leaf epidermis of willows growing in a shaded, cool environment is normally less than that of those growing in the warmer, more exposed habitat. The thickness of the leaf epidermis might be a critical barrier to penetration for a tiny gallmaker attempting to deposit its eggs. Thus, the insect must locate not only the right species of tree or shrub but also one in the right physiological condition suitable for egg deposition.

Because some insects are small and wingless or weak flyers, the location of potential host plants and the influence of the wind become important factors in the ultimate success of these organisms. The location of an oak in relation to topography and other trees influences the total population of gall organisms to be found on the individual tree. Isolated trees and shrubs and those at the top of a hill or at the edge of a woodland or brushland generally support more gall organisms and galls than plants of the same species in other ecological situations. While this cannot be considered a rule-of-thumb, proven 99% of the time, the oaks found at the top of a hill or at the edge of a woodland catch the light-bodied insects that are blown off course or away from the tree from which they were reared. Wind shelter develops within a grove as the outer trees face the brunt of the breezes filtering out organisms on the wing. The diversity and population size of individual species of gall organisms tend to be less in the middle of a grove. The dispersion of those species that do occur tends to develop at a slower rate in the center of a thick woodland.

In autumn many galls drop to the ground where the insects spend the winter in a prepupal form. Many gall inducers emerge in the spring, timed with the development of new shoots and leaves. Year after year a species of gall organism may utilize the same host tree or shrub, providing some ecological factor, such as wind, does not develop at the time of emergence to interrupt their course. In addition, the size of the population of a species varies from season to season and year to year. Unusually harsh winters with extreme temperatures may affect not only the condition of the host tree but also the number of gall organisms that survive. Assuming the insect survives the winter and successfully emerges in search of a mate and/or an egg deposition site, it must face the rigors of living in an environment fraught with new dangers. Should winds blow the insects away from the plants they require, they may perish without having reproduced. Other factors such as competition for nesting and feeding habitat, predation, and parasitism play an extremely vital role in the complex life cycles of many gall organisms.

THE GALL COMMUNITY

AT FIRST it might seem that gall organisms live a rather sheltered life, isolated from adverse elements. This notion could not be further from the truth. Actually, gall organisms are affected by the activities of a broad variety of other animals. Gall makers and their galls are the focal points upon which a number of other animals depend for survival. Galls are the foundation for an interplay of other animals such as insects, birds, and mammals. Each takes advantage of either the presence of the gall or the gall organism inside. The story of galls is one of a number that involves organisms large and small that are interrelated in a microcommunity.

The Story of Galls, Bees, and Ants

Some gall inducers release enzymes that convert plant starch into soluble sugar. In the case of several gall wasps a sweet phloem exudate develops at the surface of their galls. This material has been referred to as **honeydew.** The presence of copious quantities of this sweet material attracts ants, bees, and other insects. The cynipid wasp *Disholcaspis eldoradensis* induces pea-sized galls that exude a continuous supply of the sweet material while the galls are developing. When several hundred of these galls occur on the same oak, the tree hums with the sounds of buzzing bees. It has been reported that in a large grove of valley oaks in the San Joaquin Valley, bees gather an estimated 30-40 pounds of this "honey" in a single season. However, the high mineral content of this exudate is such that it leads to the early death of the bees. Columns of ants marching up a tree trunk may signify their attempt to benefit from the honeydew gall makers. While the ants are attracted by the sweetness, the gall wasp may benefit from the ants. Dozens of ants swarming over the galls and collecting the sweet exudate may discourage predaceous and parasitic insects from attacking the gall wasp.

Parasites and Others

A complement of mites, midges, and wasps interferes with the development of many gall organisms. Some of these organisms feed on the gall tissue, and others feed on the gall organism itself. Those individuals that feed on the gall rather than the gall organism are called **inquilines.** Though they invade plant tissues, inquilines might interfere with the gall insect in food competition.

On encountering the gall inducer, some inquilines will kill and eat it as well as the gall tissue. Occasionally, inquilines interfere with the development of the gall through their voracious feeding activities. Inquilines, such as some mites and insects, are incapable of producing their own galls, yet many have evolved a dependence on galls for food and shelter. Ironically, their survival depends on the survival of the gall-inducing organisms.

Other insects enter the gall for the express purpose of preying on its occupant. Again mites, midges, and wasps are involved. When insects begin emerging from galls placed in jars, it is difficult, at first, to distinguish the inquilines, parasites, and gall makers. In the case of wasps and oak galls, the parasites are usually metallic shades of green, blue, or bronze. Gall-inducing wasps are normally black or brown.

The degree of parasitism within a particular population of galls can be quite high. In some cases, as much as 70% or more of the population of a single species can be parasitized. In view of the degree of parasitism and all the other factors gall organisms confront, one might question how any survive at all. The fact remains that they do indeed survive. One of nature's eternal struggles unfolds as enough of each species survives to procreate from one season to the next.

The fleshy galls of aphids and some midges, which contain several young within a common cavity, are particularly susceptible to parasitic attack. A considerable number of midges attack gall-inducing organisms and borrow their homes rather than build or induce their own. Likewise, parasitic wasps such as ichneumons, braconids, and chalcids often attack other wasps within galls without discrimination. The parasitic wasp might attack the larva or pupa of the gall maker, an inquiline, or a parasite, including its own kind. Even in this event, there is no guarantee that the voracious larva that develops on the energy of its predecessor will survive to maturity. In nature's ever-fascinating way, the victor may fall victim to yet another parasite.

This seemingly complex interplay of host, inquiline, and

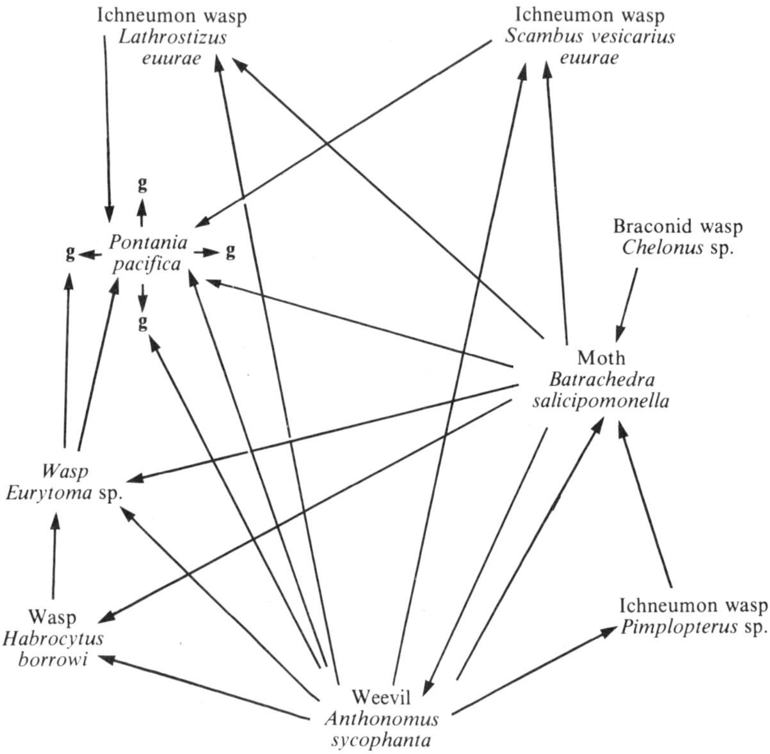

Fig. 16. Food interrelationships between parasites and inquilines associated with the galls induced by the willow-apple sawfly, *Pontania pacifica.* **g** represents plant tissue comprising the gall; arrows point to the organism or tissue killed and consumed.

Two species of wasps *Lathrostizus* and *Scambus* act as direct parasites on the sawfly *Pontania pacifica.* The wasp, *Eurytoma* sp., the moth, *Batrachedra* and the weevil, *Anthonomus* are drawn to the gall tissue as their primary source of nourishment, but indirectly act as parasites in that they will kill and eat the sawfly upon contact. In turn, each of these three insects is parasitized by wasps. However, it is interesting to note that the eggs of *Eurytoma* are laid only after the sawfly has been stung. Also of importance are the facts that the adult and larval weevils eat the gall tissue and soon enter the gall chamber where they kill and eat any insect present. The moth larva will also kill and eat any insect contacted including its own kind. When more than one moth larva occupies the same gall, normal development takes place until they contact each other; this results in one or neither of them surviving the internecine action.

parasite is best seen in the example of the willow apple sawfly, *Pontania pacifica.* At least eight different species of insects act as inquilines or parasites in the galls of this species. These include six wasps, a moth, and a weevil. The moth and the weevil are inquilines that feed mainly on the gall tissue. However, the moth larva will kill any other organism it comes into contact with, including its own kind. The same apparently is true for the weevil. Neither of these set out to destroy other insects, but do so by chance confrontation and possibly as a means of reducing competition or gaining additional nutrition. The adult parasitic wasps probe with their long **ovipositors,** which have built-in heat receptors designed to detect metabolic heat, in search of potential hosts for their eggs.

Oak Moths and Gall Wasps

Many gall wasps have emerged in search of fresh spring leaves to deposit their eggs within only to find no leaves available. The culprit is generally the California oak moth, *Phryganidia californica.* The larvas of this moth have a seemingly insatiable appetite for succulent oak leaves. During the spring and summer months of certain years when the oak-worm population is at outbreak levels, whole stands of oaks are barren of leaves. The oaks are not killed, but only pruned. They regenerate new leaves, but this takes some time, particularly if it is late in the growing season. Populations of gall wasps that depend on the presence of coast live oak or valley oak leaves to deposit their eggs can suffer serious setbacks when the leaves are either being eaten rapidly or have already been taken by the moth larvas. Fortunately, the oak worm, the oak trees, and the gall wasps have been living in association with each other for thousands of years and continue to survive.

Birds, Mammals, and the Like

Although of lesser influence than parasitic mites, midges, and wasps, larger animals such as birds, mice, squirrels, and others

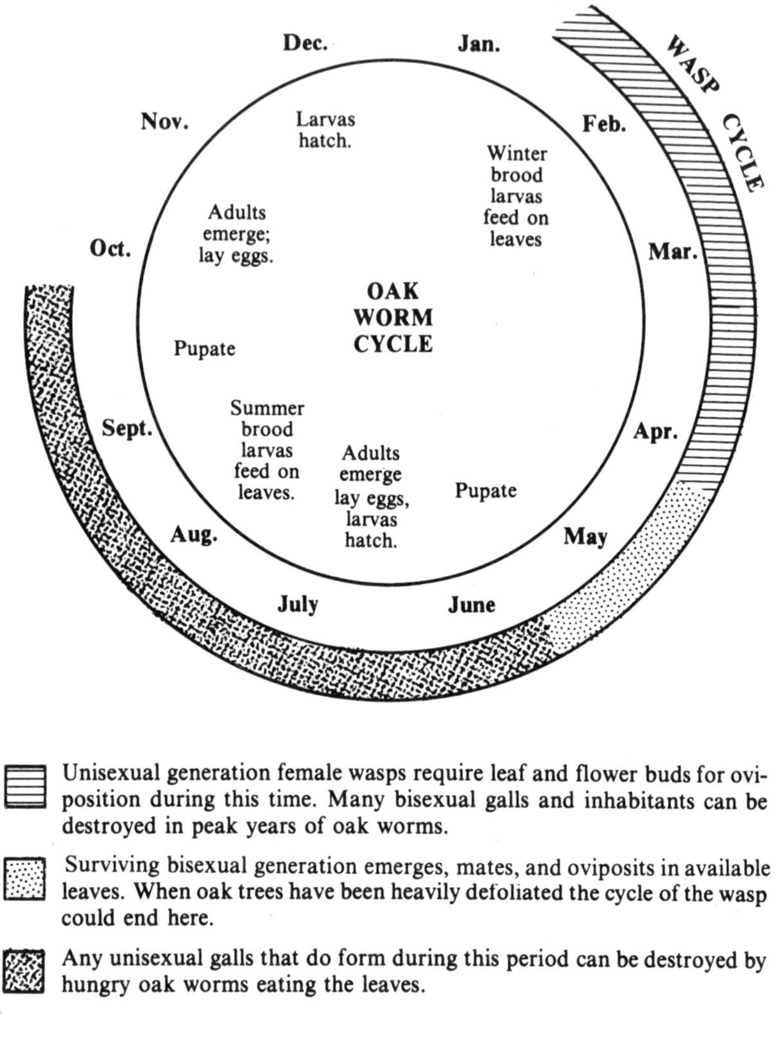

Fig. 18. **Competition between the oak worm,** *Phryganidia californica,* **and the gall wasp,** *Dryocosmus dubiosus,* **on coast live oak.**

Unisexual generation wasps emerge from mid-January through April in search of leaf and flower buds. Normally, they are able to locate suitable egg-laying sites such as these. However, in years of heavy infestation by oak worms, the wasps may be unable to locate the plant parts they require for oviposition. If they are successful, then the galls of the bisexual generation may never complete development because the hungry oak worms consume tender oak leaf tissue. Later in the spring, as the defoliation of the oaks becomes more noticeable, the bisexual adult wasps that survive may have trouble locating the tissue they require for oviposition.

As the summer brood of oak worms develops, making greater demands on the available supply of oak leaves, the unisexual generation of these wasps could be drastically reduced, if not eliminated, in peak oak worm years. The general population of these wasps in given localities follows the rise and fall of oak worm populations over several years.

Peak populations of these cynipid wasps exerts pressure on the development of the oak worms by reducing available leaf forage during the summer. During this period, the unisexual generation galls and the actions of their inhabitants cause browning and death of leaf tissue beyond the galls, thus reducing food for the oak worms.

Fig. 19. **Acorn woodpeckers** regularly excavate oak apples induced by the California gallfly, *Andricus californicus,* in search of these insects and others that are attracted to the galls as parasites or inquilines. Old galls are typically rich with a variety of insects. (Pam Olhausen)

consume some gall organisms and their galls. While scrub and Steller's jays are known to occasionally eat small galls, woodpeckers are by far the prime avian gall-insect predator.

Woodpeckers frequently dig into large galls induced by the California gallfly, *Andricus californicus.* During the fall and winter months when the tissue of these galls is soft and full of pupas or adults, woodpeckers are frequently seen chipping out holes in search of the hidden morsels. If there is a large cluster of these galls with conical holes up to 2.5 cm in diameter at the surface of the galls, that is the work of hungry woodpeckers.

When galls drop to the ground, they become available to mice and other rodents and become part of their diets. Several dozen extremely hard, spiny, sea-urchin-like galls caused by *Antron echinus,* a cynipid wasp, were left in an open container in a storage room adjoining my office a few years back. This opportunity proved too much of a temptation for the hungry mice that shared the space. Every gall in the container was chewed to pieces. These galls appeared to be the least palatable of those in our collection because of the prominence of the spines and the high tannic acid content, yet they were readily consumed.

The life of gall organisms is anything but simple. If they survive their youthful days within the galls without being parasitized or otherwise eaten by birds or mice, they may emerge as adults to face being captured by hungry birds such as chickadees, kinglets, bushtits, and vireos. Camouflaged crab spiders and dragonflies also take their toll of defenseless gall insects as they attempt to reach their host plants.

Galls that are not eaten by rodents, birds, or insects ultimately decay, releasing locked-in nutrients for recycling.

PLANTS THAT ARE GALLED

GALLS are known to occur on a broad variety of plants, including algae, mosses, lichens, fungi, ferns, conifers, and flowering species. The frequency of gall formation on algae, lichens, mosses, and fungi that occur locally is not known because little attention has been directed toward these relationships. In Europe a scientist has noted certain galls caused by parasitic fungi growing on various fleshy mushrooms, which also happen to be quite common in California. In all of the years that I have collected mushrooms for gastronomic use, I have only once found a mushroom that had a deformity that might have been a gall.

Most of the galls found in California occur on native shrubs and trees. A few galls, however, appear on ornamental plants such as oleander, azalea, peach, and olive trees. These will be mentioned later in the section on plant damage.

Gall organisms are known to occur on a few members of several diverse plant families. The largest percentage of galls is found on members of the beech, rose, willow, and sunflower families. Almost half of the known galls in America occur on oak trees, which are in the Beech Family, Fagaceae. The majority of shrubs hosts less than a few different species of gall organisms per shrub species. Such is not the case with trees like oaks, willows, and junipers. Fewer species of gall organisms appear on maples, walnuts, pines, aspen, cottonwoods, and alders. The shrubs included in the discussions of gall species in this work are manzanita, sage, sagebrush, saltbrush, thimbleberry, wild rose, creosote bush, poison oak, spiraea, chinquapin, wild cherry and plums, and coyote brush, or *Baccharis*.

DAMAGE TO HOST PLANTS

GENERALLY, the overall vigor of a shrub or tree is not significantly affected by the seasonal production of galls. Localized damage may be sustained if a branch, group of leaves, or a single leaf produces either a large number of galls or galls of a particularly large size.

Significant damage to the host plant as a result of gall production depends on several factors. Variables include the size of the plant, the conditions under which it is growing, and the degree of infestation by a particular gall organism. Severe damage can occur to an already weakened tree if it is forced to produce large numbers of giant oak apples consistently over successive years. In cases like this there is little that can be done to save the tree from further damage.

Fig. 22. A small valley oak is heavily galled by *Andricus californicus.*

When one finds an old valley oak with thousands of baseball-sized oak apples from the wasp *Andricus californicus,* one might think that the tree is under severe attack and seriously stressed. Actually the galls of this wasp may remain on the tree several years after the wasps have left. Thus, what is seen is probably an accumulation of galls representing several different generations of the same species. This does not necessarily mean that the tree is weakened and dying. While localized damage can occur to a branch, the overall vigor and health of the tree would not be endangered unless the optimum conditions for the wasp prevailed for several years and the galls were of a damaging type. Many galls cause little or no significant damage regardless of the quantity because they are neither integral, which might interrupt the flow of nutrients, nor are they located in a place where damage could be sustained. Annual changes in climate, parasitism, predation, competition, and availability of the appropriate egg-laying habitat keep the gall organisms in check and modify the stress placed on host plants.

Occasionally coast live oaks bear leaves that have the appearance of browning-out or dying in midsummer. This condition, called **scorch,** is the result of the presence and activities of the tiny wasp *Dyrocosmus dubiosus*. The unisexual generation of galls yielding this wasp forms on the feeder veins on the undersides of the leaves. A few of these galls per leaf results in only marginal browning of the leaf tissue; but when a leaf produces several of these galls on each vein, the leaf, or a portion of it, dies before the end of summer. By developing on the feeder veins, the galls interrupt the flow of essential nutrients to outer leaf tissue.

In another case the egg-depositing activities of a cynipid wasp pave the way for a plant disease that annoys thousands of people in the interior valleys of central and northern California. In the late summer a sticky ooze drips from live oak trees, covering patios and walkways. At first the ooze resembles the froth of spittle bugs (Cercopidae). Investigation has shown that the cause is actually a bacterium, *Erwinia* sp., which causes the acorns to produce a noticeable flow of sap.

The bacterium enters the young acorns through the punctures made by the egg-laying adult wasps. What effect the bacterium has on the developing wasp larva is not known. However, acorns infected with this **drippy nut disease** usually lose their viability and fail to germinate. Galled acorns rarely develop anyway, so the impact of the bacterium on the acorn crop is debatable. This seems to be a situation in which the otherwise scarcely noticed activities of a small wasp open the doors for a pathogen that seriously annoys people.

In nature these kinds of relationships have existed for many years. In most cases some sense of balance has been achieved between host and parasite. Where this balance has not been achieved, human interest is often aroused by the unsightly results or the economic loss involved. The galls that occur on ornamental plants will be discussed here because of the kind of damage these galls cause.

The leaves of peaches and nectarines are susceptible to a fungus, *Taphrina deformans,* which causes a serious disease known as **peach leaf curl.** This fungus also attacks the flowers and fruits of these trees. Spring leaves become severely swollen, distorted, and red in color. Because of the extensive cell multiplication induced by this fungus, it should be clearly recognized as a gall-inducing organism in California, as it is in England.

A similar leaf gall is formed by another fungus, *Exobasidium vaccinii,* on garden azaleas. Close inspection of the leaves of an infected plant in the spring reveals a white bloom of fungus spores on the swollen, light green new growth. Both peach leaf curl and the blight to azaleas can be treated.

Orchardists and nurserymen are quite familiar with a condition known as **crown gall,** caused by the bacterium *Agrobacterium tumafaciens.* A 1963 estimate indicated that losses caused by crown gall exceeded $6 million annually in California orchards. The bacterium is spread throughout the soil. The galls, which form just below the soil surface, often exceed a foot in diameter. Crown gall damages fruit, nut, shade, and other trees by impairing the flow of essential nutrients and by acting as a reservoir for other fungi. Secondary fungi invading a tree with crown gall can cause heartrot disease. Infections of this nature weaken the supportive ability of the heartwood, which allows the trees to topple over.

The damage caused by peach leaf curl and crown gall is considerable. However, one must keep in mind that most other gall organisms cause little, if any, significant damage to trees and shrubs. Some people have asked how they might get more of the "red and purple things" to grow on the leaves of their backyard oaks. This is an interesting idea for anyone who appreciates the beauty of the strange and colorful galls that might occur around his home and garden, rather than an approach to them as some infectious disease to be dealth with through pruning or chemical application.

GALL-INDUCING ORGANISMS

WHILE gall structures are fascinating in themselves, what about the organisms that stimulate their development?

The list of gall organisms known to induce galls is quite impressive. There are well over 2,000 gall organisms known to occur on American plants. There are hundreds of other organisms that are associated with gall development that have not been discovered, described, or identified.

Among the cast of so-called "gall makers" are few forms of plant life and a broad variety of animal life. Fungi, bacteria, and mistletoe are the most prominent plants involved in gall making. Viruses probably cause a number of galls, but their isolation and identification are difficult. The bulk of galls is initiated by small animals such as rotifers, copepods, mites, and larger creatures such as roundworms, thrips, aphids, scales, beetles, moths, midges, and wasps. The mites, midges, and wasps are responsible for more than half of the known galls in this country. In fact, almost half of the galls described in this book are caused by wasps alone.

While it is not possible to pinpoint the life cycle of every species, a look at a general description of each group will help in understanding the basic mechanisms involved in the intriguing lives of these rarely seen organisms.

Fungi

OF ALL THE PLANTS that induce galls, **fungi** are the most conspicuous. Unlike trees and shrubs, fungi lack chlorophyll and cannot produce any of their own food. Instead, fungi depend entirely on living or dead plants and animals for nourishment. Reproduction is accomplished primarily by means of **microscopic spores.** Fungus spores are frequently blown about by winds and can survive for years at high altitudes. If a viable spore lands in the right place (i.e., on the right host or substrate), it may germinate at the proper time to produce tiny fungus threads. These threads, called **hyphae,** draw nourishment from dead leaves or wood (**saprophytic**) or from living tissues of leaves, branches, or catkins (**parasitic**). The presence and activities of certain types of fungi in the living tissues of plants cause them to form galls.

While a number of different types of fungi are responsible for galls, only two groups are associated with galls in this book. The first group is known as the **rust fungi,** Order Uredinales, Class Basidiomycetes. Rust fungi are named because of the bright orange spores that form at the surface of the host plant's leaves or branches. The second group belongs to a class known as the **sac fungi,** Order Taphrinales, Class Ascomycetes. This group gets its name from the sacs that bear the spores. Not all of the rust or sac fungi induce galls. The rust fungi, for example, are responsible for a broad variety of diseases that affect the flowers, fruits, and leaves of ornamental shrubs and trees, among other plants, without producing galls.

Fungi have extremely complex life cycles. Some gall fungi require two different host plants to complete their cycle. Unlike most gall-inducing organisms, some gall fungi affect different parts of the host plant at the same time. Also, some types of gall fungi can induce galls on several different species within a genus, such as the pines.

The rust fungi are responsible for some of the largest known galls in California. Rust galls on pine trees can exceed a foot in diameter. These large galls take several years to mature, since their growth is slow. Some rust fungi cause witches' broom galls. These broom galls are composed of compact clusters of shoots that arise from a common base on stems and branches. The types of rust galls most often found are induced by the fungus *Peridermium* on pines, and the fungus *Puccinia* on coyote brush or *Baccharis.*

Sac fungi of the Order Taphrinales produce spores that survive on various parts of the host plants during winter months. Young, tender parts of a plant become infected in the spring or early summer. The galls induced by these fungi occur on the leaves and catkins of certain plants. The fungus *Taphrina deformans* was mentioned earlier in connection with peach leaf curl, and from an economic point of view is the most important of the gall-forming fungi in the class Ascomycetes. However, another species of *Taphrina* induces flamboyant, colorful, tongue-shaped galls on the catkins of alder trees. Other species of *Taphrina* not described herein cause swelling and distortion of leaf tissue and young, green cones on alders. Another ascomycete fungus, *Dibotryon* sp., causes black knot galls on the branches of choke cherry.

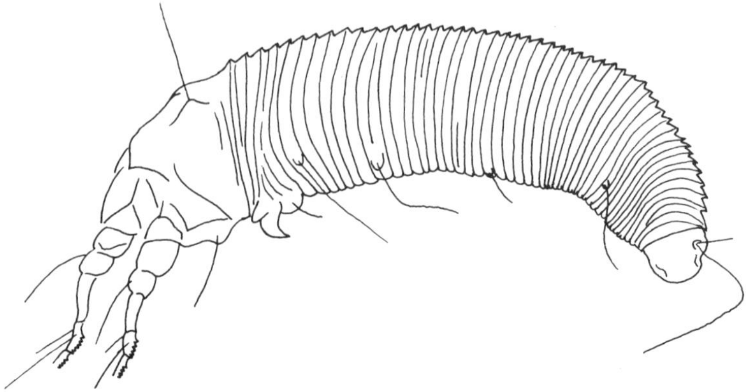

Fig. 28. Eriophyid mite, *Phytoptus emarginatae*
Actual size .3 mm.

MITES

Order Acarina
Family Eriophyidae

MITES are fascinating animals distantly related to spiders and scorpions. Some mites are microscopic. For years they have been recognized as serious pests of livestock, vegetable, and fruit crops. These tiny animals occupy a broad variety of ecological niches from walnut leaves to pig skin. One of the most interesting groups of all the mites belongs to the Family Eriophyidae, which is thought to have originated more than 50 million years ago.

A number of eriophyid mites are intimately associated with plants in gall-forming relationships. Eriophyid mites are tiny, elongate creatures that average about 0.2 mm in length. With their piercing mouth parts, eriophyid mites are able to drain the contents of individual plant cells.

A single egg-laden female can initiate the formation of a gall through feeding and piercing and wounding the plant as well as releasing chemicals into the wounds during the process. Some females who do not come into contact with **spermatophores,** the stalked sacs containing sperm from males, produce only males. Those females that do contact spermatophores produce males and females. In some cases females retain sperm through the winter after contacting male spermatophores. In the spring these females produce offspring of both sexes. The males do not survive the winter.

Offspring remain in the gall of the parent and contribute to further gall expansion through their feeding activities. Young mites pass through two nymph stages. After a resting period, the adult usually develops from the second nymph stage. Growth from egg to adult sometimes takes only two weeks. Adult females often hibernate in the fissures of bark and among bud scales. In the spring and early summer adult females emerge to infest new leaves, flowers, and fruit. Mites travel by drifting in the wind, riding on insects and birds that contact host plants. This hitchiking technique assists in dispersion of the mites onto noninfested host plants.

Each species of mite is normally found on a specific type of tree or shrub. Like some of the fungi, however, some mites are found on several species of plants within a generic group, such as *Eriophyes mackiei* on black oaks and canyon oaks.

Eriophyid mites initiate several different types of unusual galls, some of which are unique to mites. Eriophyids induce flower, bud, leaf roll, pouch, and erineum pocket galls. Most mite galls are not very colorful nor as flamboyant in design as midge and wasp galls. Mite galls are characterized by an outer covering of minute, fuzzy hairs and a size usually less than 1 cm. Many **pouch galls,** which have microscopic holes or openings at the base, are not notably hairy. Others are sparsely covered with short hairs. While pouch galls assume a variety of shapes, they are normally asymmetrical, lopsided, club-shaped growths. The beadlike galls found on alder and willow leaves are actually stubby pouch galls.

Erineum pockets develop as hair-covered depressions, usually on the undersides of leaves. The mite *Eriophyes mackiei* feeds on the undersides of the leaves of black and canyon live oaks. The erineum pockets that result are rust colored and may cover the entire undersurface of the leaf. The upper surface of the infested leaves displays corresponding lumps that match each erineum pocket below. Another eriophyid induces light yellow erineum galls on walnut trees. A third eriophyid induces globular pouch galls on the upper surface of walnut leaves.

Eriophyid galls are commonly found on the leaves of alder, willow, poison oak, bitter cherry, choke cherry, sierra plum, oak, walnut, and maple.

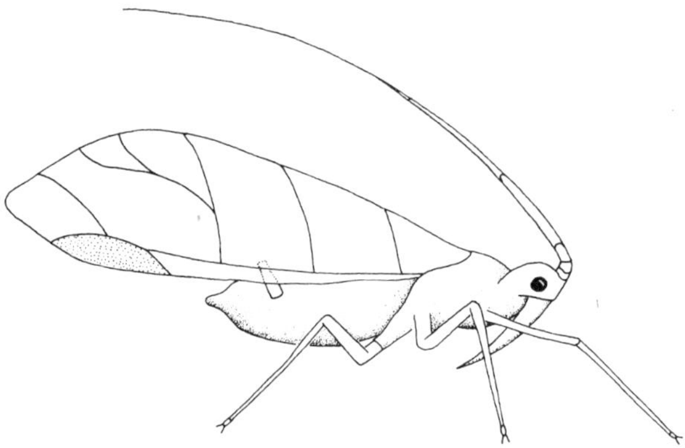

Fig. 30. Aphid, a winged alate
Actual size 2 mm.

APHIDS

Order Homoptera
Family Pemphigidae

APHIDS belong to the same order of insects as leafhoppers, spittlebugs, mealybugs, and scale insects. In one way or another, these insect groups have been long known for their damaging effects on commercial crops and fruit.

Aphids feed by piercing the epidermis of a leaf or bud with their long needlelike mouth parts (**stylets**) and sucking juice from the phloem tissue. Aphids are capable of converting plant starch into soluble sugar by releasing enzymes. Because of their method of feeding, aphids are known to be important vectors of viral plant diseases. In some cases, viruses are taken into the salivary glands and later retransmitted into plant tissue through the stylets.

Like other gall organisms, the life cycles of aphids are extremely complex and not well understood. While we cannot apply a single life cycle description to all aphids, there are certain characteristics that are generally applicable.

Some gall-forming aphids exhibit an alternation of generations. In the fall the female of the sexual generation deposits an egg under the bark of a tree. In the spring the egg hatches, releasing a wingless female called a **stem mother.** With other gall aphids stem mothers are capable of going through several generations throughout the year, parthenogenetically, without an egg stage. Continuous parthenogenetic reproduction without reference to an egg stage is not uncommon among many species of aphids.

In the case of the gall aphids the stem mother initiates the gall through her feeding behavior. After reproducing parthenogenetically, the stem mother shares the gall with her wingless female offspring. Their combined feeding activity probably stimulates further expansion of the gall. Some gall aphids are covered with a filamentous, waxy, white material produced by special secreting organs. This material is often seen in the galls of *Pemphigus* aphids. It may function in protecting the aphids from certain intruders.

The offspring of the stem mother moult several times to eventually become **alates.** In some cases these alate females reproduce parthenogenetically, producing the sexual generation. In others the winged forms fly to an alternate host. In the fall a special group of these alate females fly back to the primary host where they give birth to the males and females. The alternation between two separate plant hosts is another feature that makes aphids somewhat unique. Some aphids require two host plants, and others get along on a single species. Some of the most damaging aphids, economically speaking, attack sugar beets and lettuce during part of their cycle, while spending the rest of their lives on alternate hosts.

The galls generated by aphids are among the most highly specialized of insect galls. Most of the galls induced by aphids are of the roll, fold, or pouch type on leaves or leaf-bearing stems (**petioles**). One of the aphid galls appears as an asymmetrically formed bulge with a tightly closed slit on the lower surface. The aphids leave the gall after it matures, expands, and a slit opens allowing their escape. This usually coincides with the changes in the plant's physiology in the fall months.

Gall aphids are commonly associated with cottonwoods and manzanita. Aphids of the genus *Pemphigus* appear on cottonwoods, while aphids of the genus *Tamalia* appear on manzanita. The taxonomy of *Pemphigus* spp. is in an extreme state of flux. They appear to be a terribly confusing group, leaving much research to be done in order to carefully separate the species and identify new forms. In this work I have chosen to use species names that best fit the galls I have studied, with the full awareness that they are subject to change with further research. I have placed a question mark after each such species designation, indicating that such designations may prove incorrect with more study on the part of many researchers. For the time being, these taxonomic usages are the best that can be produced.

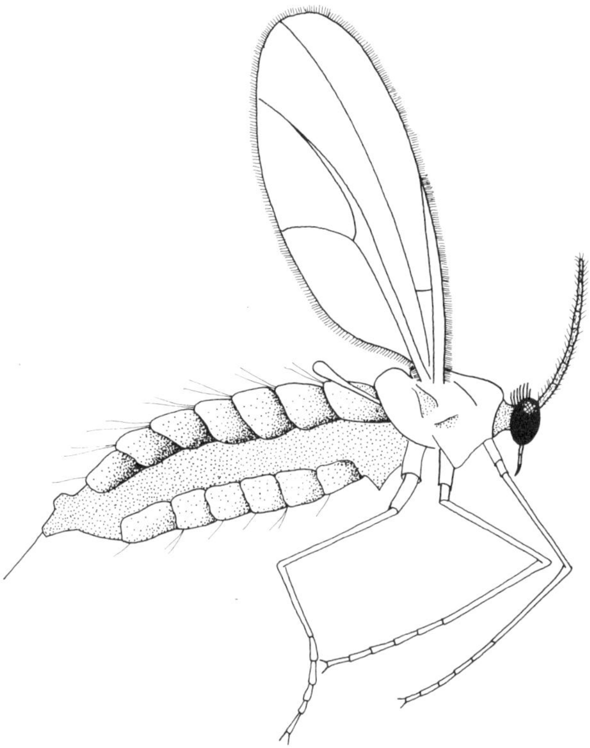

Fig. 33. Cecidomyiid midge, *Asphondylia* sp.
Actual size 5 mm

MIDGES

Order Diptera
Family Cecidomyiidae

OF THE three families of flies known to induce galls, the family Cecidomyiidae, the **gall midges,** is the most important and dominant. While the larvae of some cecidomyids feed on fungi and forest debris, or prey on mites and insects, most cecidomyiids feed on plants, and induce some of the more fascinating galls.

Some gall midges, like other insects, have an alternation of generations with a sexual generation of males and females alternating with a unisexual generation of females. With these species the galls of one generation form on different parts of the plant from galls of the same species of the preceding generation. Most gall midges do not have alternating generations. Females deposit their eggs on appropriate plants following mating. The larvas initiate the formation of the galls through their feeding activities. With some species of gall midges, the gall contains a single larva (**monothalamous**). Within the galls of other species, several larvas may occupy a single chamber (also monothalamous) or the same gall but with each larva in its own chamber (**polythalamous**).

Gall midges may be divided into two artificial groups: those that pupate within the gall and those that pupate outside of the gall. Some gall midges pupate within silken cocoons among leaf litter and humus on the ground. Other gall-forming midges pupate on the outside of the gall on its surface. Most adult cecidomyiids are able to feed on nectar during their short lives. Their primary function as an adult is to reproduce and continue their species.

Midge galls exhibit an enormous diversity in form. Cecidomyiids initiate a variety of pouch, roll, fold, petiole, leaf, and bud galls. The greatest majority of midge galls are found on leaves, with fewer numbers occuring on buds, stems, and flowers. Leaf, fold, and roll galls induced by midges differ from those of mites and other insects in that they appear much more brittle. Midge fold and roll galls are difficult, if not impossible, to open without breaking the galls. This is clearly seen in the case of the spiraea pod gall caused by *Rhabdophaga salicifolia*. The swollen flesh of this gall is 2-3 mm thick and extremely brittle.

Midge galls occur on a broad variety of plants, including sage, creosote bush, sagebrush, saltbush, spiraea, coyote brush, willow, and juniper. The genera of cecidomyiids commonly encountered include *Oligotrophus* and *Walshomyia* on juniper and cypress; *Rhabdophaga* on willow and spiraea; *Rhopalomyia* on coyote brush, sage, and sagebrush; *Asphondylia* on creosote bush, silk tassel, and saltbush; and *Dasineura* and *Neolasioptera* on lupine.

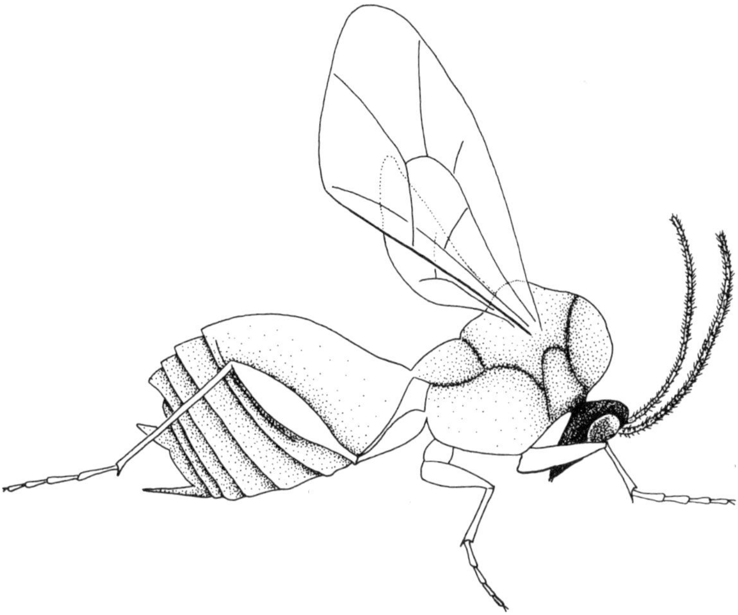

Fig. 35. Cynipid wasp, *Andricus* sp.
Actual size 4 mm

WASPS

Order Hymenoptera
Families Tenthredinidae and Cynipidae

UNLIKE THE larger wasps, which seem to specialize in plaguing those attending barbeque parties and camping trips, the gall wasps are small insects whose life habits are neither threatening nor aggressive. Of all the gall insects, the wasps are the real stars of this ecological relationship. Gall wasps, particularly the cynipids, are responsible for the most brilliantly colored and oddly shaped galls to be found. While a few gall wasps approach the size of an ordinary house fly, most others are about the size of a fruit fly or smaller.

There are two primary groups of gall wasps. The first group belongs to the family Tenthredinidae and is known as **tenthredinids** or **sawflies.** These wasps have inherited this misleading name because of their sawlike ovipositors. Those sawflies that induce galls can be divided into two important genera, which are common and extremely host-specific. Members of the genus *Pontania* initiate galls on the leaves of willows throughout California. The egg-laying adult, in some cases, is largely responsible for gall formation, since the galls reach full growth before the egg inside hatches. *Pontania* galls are normally round or ovular, red or green, and have a large larval chamber. Certain members of this group cut a hole through the gall and clean house by ejecting acculumated fecal matter prior to emerging. The larvas of *Pontania* sp. emerge from the galls and drop to the ground by a silken thread. The larvas build cocoons among the leaf litter where they pupate and emerge later as adults.

The second genus of tenthredinids causes branch and petiole galls on willows. The galls that result from the actions of these wasps are either gradual or abrupt swellings. The sawflies responsible belong to the genus *Euura.* Many of the larvas of this genus also clean house in the above manner. However, most members of this group pupate within the galls. The larvas of some species cut exit holes near the top of the galls before retreating down them to spend the winter near the base of the cavities. Around January or February along the coast (later at higher elevations) they pass through the final pupal stage and emerge as adults just about the time new willow shoots are being formed. The adults of both *Pontania* and *Euura* have short lives of up to two weeks. Apparently, these sawflies have but a single generation a year.

The second major group of gall wasps belongs to the family Cynipidae. The cynipid wasps are by far the most common group of gall insects in California. They induce some of the most interesting and unusual galls that can be found. Galls that look like miniature stars, sea urchins, golf balls, cups, saucers, clubs, and teardrops are among the fascinating jewels induced by cynipid wasps.

Unlike sawflies, cynipid wasps occasionally exhibit two

different generations per year. An alternation between a spring bisexual generation and a summer/fall unisexual generation results in the development of a different type of gall on different parts of the host plant for each generation. This has made investigation extremely difficult and has been the basis for much confusion, misidentification, and the naming of new species erroneously. The life cycles of most cynipid wasps have yet to be worked out. In many species only females, which reproduce parthenogenetically, are known to exist.

The length of the life cycle of some known cynipids lasts as long as two or three years before the adult emerges. In some of the species with early spring bisexual generations the cycle from egg to adult may be complete in two or three weeks. The summer and fall unisexual female generation may take anywhere from several weeks to three or four years to complete the cycle from egg to adult.

Like aphids, cynipid larvas secrete enzymes that convert plant starch into soluble sugar. It is this sweet, converted material that the larvas feed on. Cynipid larvas do not release fecal matter into their galls as do sawflies. Instead, cynipids have a blind intestine, at the end of which the fecal matter collects during the growth of the larvas. After the larvas reach their full growth and prior to phasing into diapause or pupation, a connection with the anus develops. Clear evidence of fecal matter existing in galls containing mature pupas has been lacking in most of the galls I have examined.

The larvas of cynipid wasps initiate gall development and later pupate in the galls. They may emerge as adults in the spring or fall of the following year depending on their cycle, the geographic location, and the timing of bud and leaf development. Altitude and climate, which may effect potential host plants, probably also have some influence on the emergence of adults.

The distinguishing characteristic of cynipid galls is that they normally consist of a series of concentric rings or zones of nutritive material around the larval chamber. However, galls induced by the wasp *Trichoteras vaccinifoliae* are the exception to this rule in that they are thin-skinned and hollow with radiating fibers leading to an equally thin-skinned larval chamber. The wall of the larval chamber becomes the nutritive zone, which is rich in reserve food material and amino acids. Some cynipid galls contain only one chamber and, thus, only one larva. Other galls contain a number of larval chambers, each with its own larva. This feature further distinguishes cynipid galls from galls of aphids and some midges where several larvas occupy a common chamber.

Cynipid galls are found literally on every part of an oak, including roots, branches, leaf petioles, leaves, flowers, and fruit. Some of these galls may drop off with the leaves or flowers at the end of the growing season or persist on the branches for several years. With the tremendous abundance and diversity of oaks in California, it is not unlikely that one's first experience with galls will be with those induced by cynipid wasps. In addition to oaks, cynipid wasps gall wild rose, thimbleberry, and chinquapin. The vast majority of cynipid galls in this region, however, are found on oaks.

OTHER GALL ORGANISMS

THERE ARE two other gall organisms that are not associated with any of the other groups previously mentioned. The first is **mistletoe**, which is a parasitic plant that induces galls on the branches of pines and oaks. The galls induced by dwarf mistletoe, *Arceuthobium campylopodum*, on lodgepole pine, *Pinus murrayana*, are particularly noticeable due to the extensive swelling. The galls produced by other species of mistletoe on oaks are less noticeable. The seeds of these parasitic plants are spread by wind, birds, and squirrels.

The second gall-inducing organism is a moth that attacks the shoots of coyote brush, *Baccharis* sp., in the spring. The moth *Gnorimoschema baccharisella* induces large, symmetrical, green stem swellings. Details on both these organisms will be presented in the species accounts.

2

IDENTIFYING PLANT GALLS

LEARNING to recognize and identify plant galls is a process that requires determining the host plant by genus and in many cases by species. It is also important to be able to observe subtle characters of both the plants and galls in order to arrive at a satisfactory conclusion. Since in a few cases there is some confusion over which is a bud gall and which is a stem gall, care must be taken in examining the specimens.

For these reasons, this section will begin with a set of simplified keys to the host trees and shrubs and a key to the gall species. For a more detailed and definitive description of potential host plants, one should refer to the standard works of Phillip Munz, Roxanna Ferris, and McMinn and Maino listed in the bibliography.

Using the Botanical Keys

The following botanical keys are designed to facilitate the use of this introductory guide and are in no way intended to be comprehensive. Hopefully, the reader will encounter galled plants included in this guide.

A Note to the Reader: The keys are based on a paired comparison of characters with sets of lettered alternatives. Carefully proceed with each choice until you arrive at a satisfactory conclusion. If a conclusion is not reached, then retrace your steps to the last major category. Repeat the process, paying careful attention to each alternative and decision. If this fails, then check the species listed in the appendix and a more definitive key. The glossary explains the terminology used in the key.

SIMPLIFIED KEY TO HOST TREES

A. Deciduous trees growing near springs, wet meadows, stream and river banks

B. Leaves heart-shaped to broadly elliptic, large sharp or round-toothed margins, and pointed tips, not deeply lobed

C. Leaves with or without rounded teeth along the margins, leaf surface smooth, pale to dark glossy green above, lighter below wave easily in breezes

D. Trees with grey fissured bark **cottonwoods, poplars,** fig. 41A

DD. Trees with white-light grey, smooth, thin bark, trunks twisted, gnarled, or straight, in groves in high-altitude meadows and streams .... **aspens,** fig. 41B

CC. Shrublike to full tree forms, with deeply veined, sharp-toothed, dull, not glossy and smooth, leaves, often with last year's small brown cones .... **alders,** fig. 41C

BB. Often shrublike trees with elongate-linear or broad, lobed leaves

C. Leaves narrow-linear to oblong, glossy to dull, smooth, with or without hairs, silvery pubescence, fine-toothed margins, flowers in compact, long catkins, seeds hair-tufted, in wet places at all elevations .... **willows,** fig. 41D

CC. Leaves broad, smooth, 3-5 lobed, toothed margins, at high altitude wet places .... **mountain maple,** fig. 41E

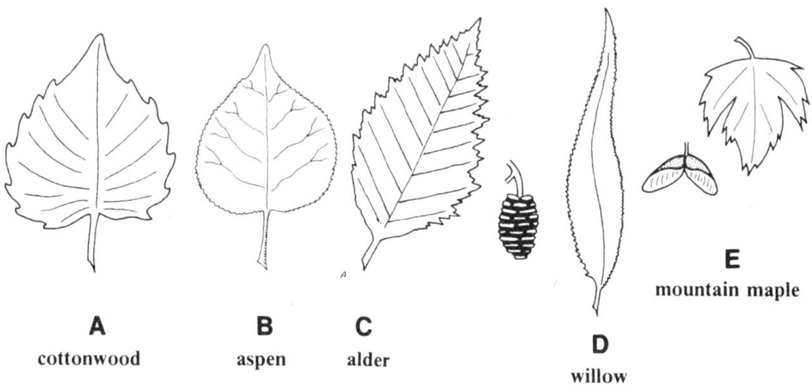

Fig. 41

AA. Evergreen or deciduous trees with shrublike forms not confined to wet places, but in open woodland, forest, or desert areas

B. Leaves scale or needlelike, evergreen, cone-bearing trees

C. Leaves small, scalelike, overall form cylindrical

D. Tree to shrub forms, cones rounded with flat to convex, nippled scales, along coast on marine terraces and mountains .... **cypress,** fig. 42A

DD. Shrublike trees in hot, dry, arid areas, tall trees at high elevations in Sierra, leaves with glandular pits, fruit a berrylike cone with a waxy coating ..... **junipers,** fig. 42B

CC. Leaves linear, thin needles, single or in bundles of two or more

D. Needles long 3-18 cm, cones often large with prickly or pointed scales, trees tall, forming forests from coast to high mountains .... **pines,** fig. 42C

DD. Needles somewhat flat, sharp pointed, single, not exceeding 4 cm, encompassing entire stem, pyramidal trees north coast counties, coastal bluffs to high mountains .... **spruce,** fig. 42D

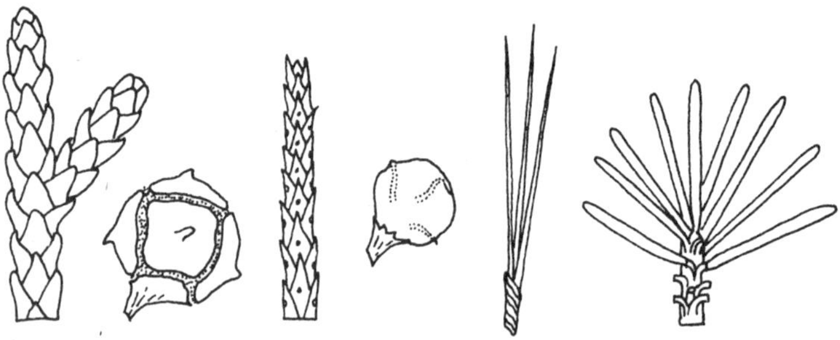

A cypress — B juniper — C pine — D spruce

Fig. 42

BB. Leaves broad, flat, evergreen or deciduous, with or without marginal spiny teeth or lobes

C. Leaves thick-skinned with teeth and/or deep lobes

D. Fruit a nut attached to a cup, flowers in loose catkins, leaves smooth, glossy to dull, along coast, dry valleys and foothills, low-elevation trees tall and broad, high-elevation forms shrublike .... **oaks,** fig. 43A

DD. Leaves broad, 3-5 lobed, with veins radiating from point where petiole meets leaf, fruits are paired, winged nuts, occurs on moist hillsides .... **maples,** fig. 43B

CC. Deciduous trees with aromatic bark and leaves, leaves large with numerous paired, opposite leaflets, fruit covered by thick, green husk, some cultivated, near residential, farmland areas, lowland countryside and foothills .. **walnuts,** fig. 43C

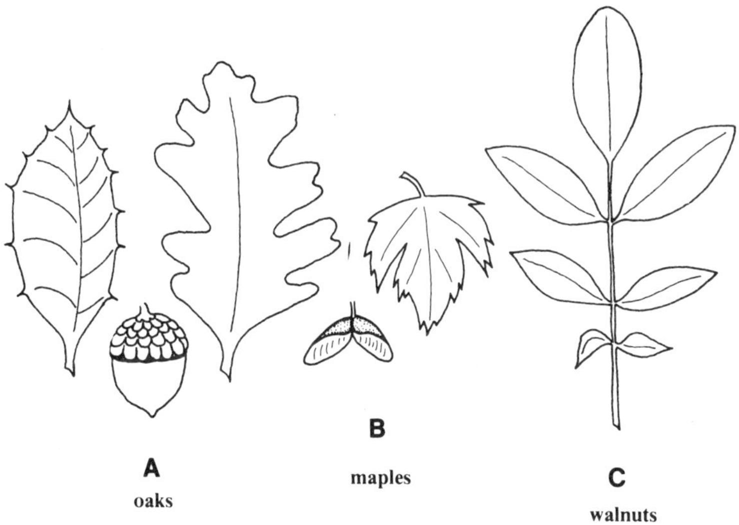

Fig. 43

SIMPLIFIED KEY TO HOST SHRUBS

A. Leaves small, generally under 1.5 cm wide, if wider then with toothed margins and spiny branches, or with smooth, red-brown, scaly bark; others with silver-grey, pubescent, aromatic leaves, or resinous aromatic leaves

B. Plants with silver-grey pubescent leaves, some aromatic

C. Leaves thin, linear-elliptic, smooth

D. Leaves spicy-aromatic, either split into linear, needlelike divisions or flattened toward distinctly broadened, lobed tips, in high Sierra and Great Basin or coastal mountains .... **sagebrush**, fig. 44A

DD. Leaves not strong aromatic, without petioles, linear to oblong, silver-grey, somewhat scaly, fruit with four toothed wings, in alkali sinks, dry flats and washes, and valleys of arid regions .... **saltbush**, fig. 44B

CC. Leaves thick, lumpy surface, toothed margins, more than twice as long as wide, grey-green, pubescent mainly beneath, glandular-aromatic, on dry slopes below 5,000 feet .... **sage**, fig. 44C

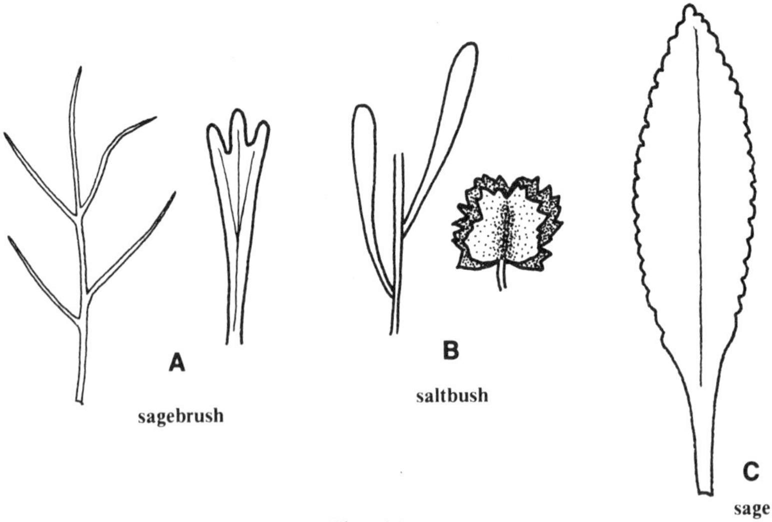

Fig. 44

BB. Leaves glossy green, smooth or toothed margins with spiny branches, or with smooth, red-brown, scaly bark

C. Leaves glossy green, smooth to scaly

D. Plants aromatic, leaves small, grouped, glossy and aromatic, stems marked with dark bands, flowers yellow, dominant shrub in many areas of Mohave Desert .... **creosote bush**, fig. 45A

DD. Leaves glossy green, toothed margins, smooth to scaly, thick, leathery, male and female flowers on separate plants, hairy seeds in autumn, grows in dense prostrate mats along immediate coast or as an erect shrub inland to Sierra foothills .... **coyote brush**, fig. 45B

CC. Leaves dull green, noticeably toothed and spiny, if smooth-edged, then without spines, dull to glossy grey-green to dark green leaves, bark smooth and red-brown

D. Thorny shrubs with dull green, ovate, toothed leaflets, pink flowers, in the open or in woodland .... **rose**, fig. 45C

DD. Leaves often larger than 1.5 cm wide, smooth edges, bark smooth-scaly, red-brown, flowers white-pink, urn-shaped .... **manzanita**, fig. 45D

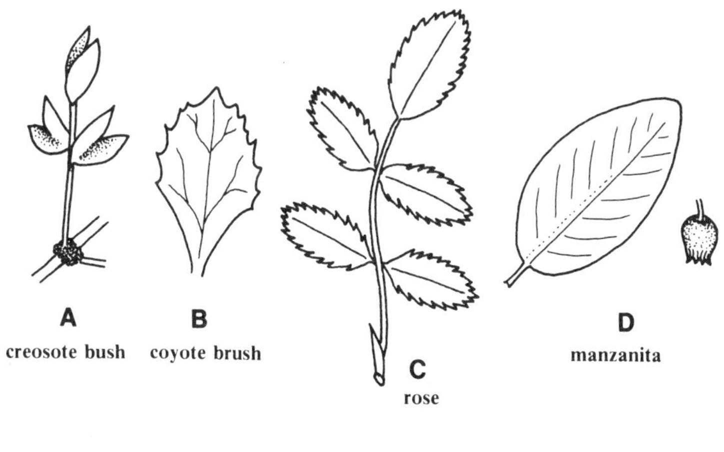

Fig. 45

AA. Leaves or leaflets larger than 1.5 cm wide, elliptic to broad lobed, thick, leathery, glossy green, or thin, dull green, or silver pubescent

B. Leaves oblong-elliptic, dark green, glossy above, pubescent beneath, smooth margins, some manzanitas fit here

C. Leaves normally more than 5 cm long

D. Leaves to 15 cm long, golden-rusty pubescent below, fruit globular covered with stiff spines, along coast to Sierra .... **chinquapin**, fig. 46A

DD. Leaves tapered at both ends, to 6 cm long, flowers single or in long, drooping catkins, fruit smooth and fleshy, in the open or among trees of coast and Sierra .... **silk tassel bush**, fig. 46B

CC. Leaves under 5 cm long, narrow-elliptic, pointed tip, glossy green above, sparse pubescent below, flower a catkin, fruit a nut attached to a cup, forming dense low-growing stands on dry, rocky slopes in high mountains north coast and Sierra .... **huckleberry oak**, fig. 46C

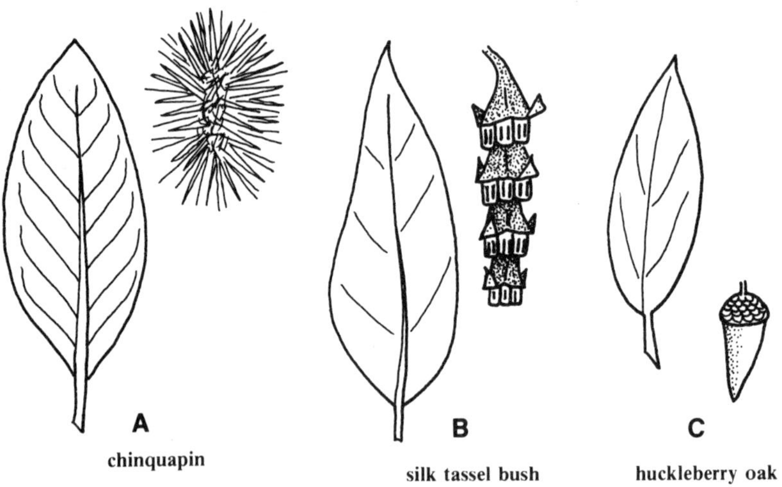

Fig. 46

BB. Leaves broad lobed or divided into leaflets of three or more, or with toothed margins

C. Leaves broad and lobed or oblong-elliptic with toothed margins

D. Leaves with fine to coarse toothed margins, but not lobed

E. Leaves with fine-toothed margins, 2-8 cm long, fruit a flesh-covered nut, purple-red, 1 cm in diameter, flowers white-pink, in damp places or rocky ridges from coast to high Sierra .... **cherries/plums**, fig. 47A

EE. Leaves round to oblong, coarse toothed along the apical half of the leaf, flowers in terminal, hemispheric, pink-rose clusters, in wet meadows and rocky areas of high Sierra and north coast counties below 11,000 feet .... **spiraea**, fig. 47B

DD. Leaves broad to 15 cm wide, soft, pubescent both sides, 3-5 lobed, pale green, light brown peeling bark, flower large and white, fruit a hemispheric red berry, in coastal woodland as tall shrub, in Sierra rarely over 0.5 m .... **thimbleberry**, fig. 47C

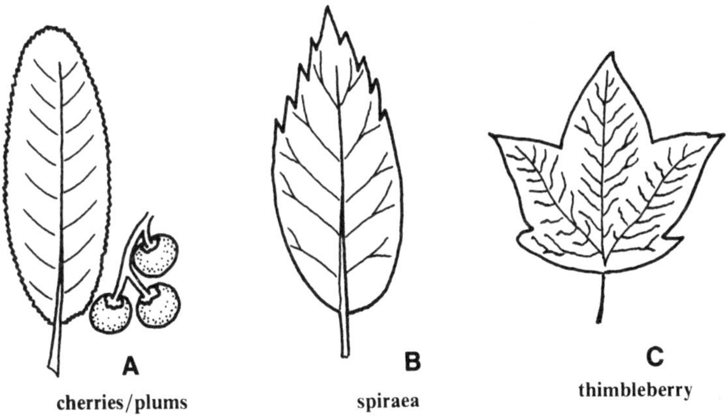

Fig. 47

CC. Leaves divided into leaflets of three or more

D. Leaves large oblong, deeply veined, margins smooth to rounded lobes, leaves bright red, orange, purple in fall, bark smooth, light brown, grows as shrub or climbing vine .... **poison oak**, fig. 48A

DD. Leaves divided into linear, narrow, silky, silver pubescent leaflets, 5-12 in number, leaflets radiate from base of petiole, flowers in tall columns, blue-yellow, seeds in long pods, along coastal bluffs and terraces .... **lupines**, fig. 48B

A
poison oak

B
lupines

Fig. 48

SELECTED GALL-HOST HABITATS

California has a diverse array of plant associations referred to as communities. Each plant community is comprised of plant species unique to that geographical region or altitude. Likewise, each plant community hosts a unique assemblage of gall organisms. The following photographs depict some of the more prominent plant communities with mention of the dominant gall organisms to be found.

Soft chaparral along the coastal ranges is composed of coyote brush, *Baccharis* sp., with small patches of California sagebrush and poison oak. Rust, mite, and midge galls are abundant in this habitat. Ultimately, chaparral-covered hills like these in the Oakland-Berkeley area are invaded by trees like the oaks and bay laurels that can be seen towering over the smaller coyote brush. (Ron Russo)

The upper, western slopes of Mt. Diablo support a broad variety of galled plants like the western junipers above. Other trees and shrubs found in the same area include canyon live oak, scrub oak, coast and interior live oak, and black sage. Several mite, midge, and wasp galls are found here. (Ron Russo)

The interior foothills of the coast and Sierra ranges are covered with old stands of blue oaks as seen here on the lower western slopes of Mt. Diablo. These trees were once considered to encourage grasses of poor nutritional quality for grazing. Thousands of blue oaks have been cut, burned, and pulled over under this belief. However, recent studies have indicated just the opposite. Thus, one of the most spectacular gall-producing trees in the western United States may receive some protection from grazing interests. Unfortunately, seedling production among blue oaks is poor, which leaves a puzzling problem for scientists while increasing the importance of protecting existing groves. (Ron Russo)

Uphill, isolated trees such as this blue oak often catch a large number of tiny wasps blown by stiff breezes during their emergence. Also, trees at the edges of groves often support more galls than those in the center of groves, particularly with oaks. Wasps are the dominant gall organisms in oak forests, with a few midges and mites making their appearance. (Ron Russo)

The rivers and streams of inner coast mountains, the great Central Valley, and the Salinas River Valley (below) support Frémont cottonwoods and arroyo and Brewer's willows along with red alders. Aphid, sawfly, fungus, and mite galls are common in these habitats. (Ron Russo)

The granitic slopes of the high Sierra, as seen here in Alpine County, support galled plants like western junipers, huckleberry oaks, Sierra plum, and pinemat manzanitas. Mite, aphid, and midge galls are common in this area. High winds and long periods of snow cover make this environment difficult in which to survive; yet, huckleberry oaks support over a dozen species of gall organisms, some of which are new to science. (Ron Russo)

High altitude meadows like this one at Bear Valley, Alpine County, support galled plants such as lodgepole pine and Spiraea. Mistletoe, rust, and midge galls abound on these plants. In addition, a variety of grasses and goldenrod plants thrive in such meadows. Our western goldenrods do not support the variety and abundance of galls that the eastern species harbor. (Ron Russo)

High mountain streams in the Sierra are characterized by Scouler's, Lemon's, and Geyer's willows, mountain alder, and black cottonwood (tall tree in the background). Sawfly galls abound on the willows, while two species of aphid galls are found on the cottonwood. In wet places along streams or roadcuts, the delicate mountain maple is often found with its mite erineum galls. (Ron Russo)

The Great Basin habitat on the eastern side of the Sierra supports such plants as the Utah junipers seen standing amid thousands of acres of sagebrush. In the background, at higher altitudes, small groves of quaking aspen thrive in spring-fed soils. The high desert community produces a variety of less spectacular galls than does the blue oak forest. The desert plants here support midge galls, while the aspens harbor aphid and mite galls. (Ron Russo)

The creosote bush scrub-shadscale-scrub mix of the Mohave Desert near Trona, California, is searched by the author and his son for new midge galls. Creosote bush is one of the most abundant shrubs in the desert. It supports several species of galls, most of which are not clearly identified by entomologists. Other desert plants like fourwing saltbush have but a single gall organism associated with their foliage. (Ken Russo)

3

USING THE GALL KEY

Now that some determination has been made regarding the identity of the host plant, one can proceed to identify the species of organisms by the galls found. In using a key such as this one, it is important to carefully weigh the sets of opposing alternatives and characters, making a decision on the basis of those characters that match the gall. Reference should be made to the drawings of the basic types of galls on page 6 in order to make the primary decisions in this key. If the reader is unable to reach a clear identification through the key and the matching description in the species accounts, then one should:

1. Go back to the major category division and repeat the process, paying close attention to any oversights in running through the key again.

2. Check the appendix. The gall may not be in the key and descriptions. This will require identification of the host plant.

3. The gall may be unlisted and unidentified, thus offering the reader the opportunity to study it, rear the organism where possible, and make a contribution to this field.

Keys to the Galls in this book

Group I—Flower, Catkin, or Fruit Galls

A. Galls club- or kernel-shaped, small, single or in clusters, on oaks

B. Galls kernel-shaped, 3-4 mm long, pointed tips, in grapelike clusters, light brown .... *Callirhytis serricornis,* fig. 56A, p. 96

BB. Galls dark chocolate brown, club-shaped, not in dense clusters, also on new spring leaves .... *Dryocosmus dubiosus,* fig. 56B, p. 96

AA. Galls round-cylindrical or fasciated, leaflike appendages, encompassing part or all of the catkin or fruit

B. Galls solid, round to cylindrical

C. Galls round, 7-13 mm single, pubescent, often at the tip of the catkin, on chinquapin .... *Dryocosmus castanopsidis,* fig. 56C p. 156

CC. Galls cylindrical

D. Galls fleshy encompassing part or all of the catkin with some floral parts protruding, to 4 cm long, on black oaks .... *Callirhytis congregata,* fig. 56D, p. 95

DD. Galls aborted fruit, spongy, round to potato-shaped, to 5 cm long, green when fresh in spring, beige and wrinkled later, on wild cherries .... *Taphrina pruni-subcordatae,* fig. 57A, p. 179

BB. Galls fasciated outgrowths of floral parts or rounded with prominent slit or opening

C. Galls fasciated outgrowths

D. Galls leaflike, knobby, red; poison oak **mite,** fig. 57B, p. 166

DD. Galls elongate, tonguelike projections of cone scales, on alders .... *Taphrina occidentalis,* fig. 57C, p. 78

CC. Galls rounded, fleshy with obvious slit, on manzanita .... **aphid**, fig. 57D, p. 165

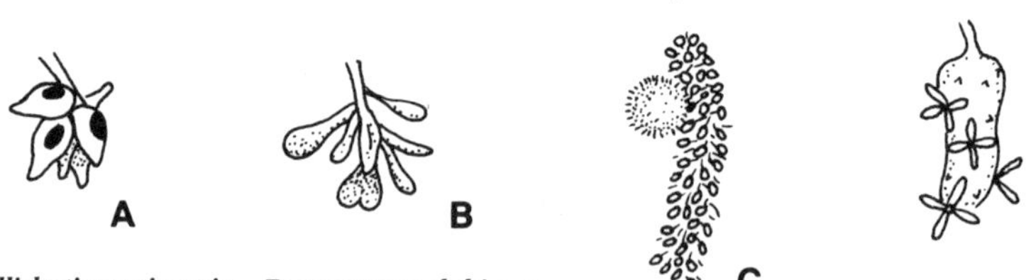

A *Callirhytis serricornis* **B** *Dryocosmus dubiosus* **C** *Dryocosmus castanopsidis* **D** *Callirhytis congregata*

Fig. 56

A
Taphrina pruni-subcordatae

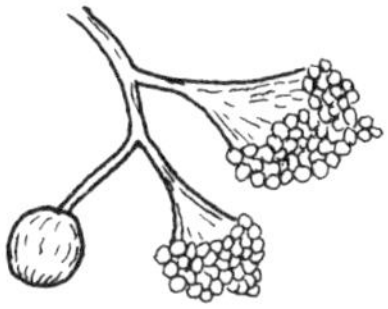

B
poison oak mite gall

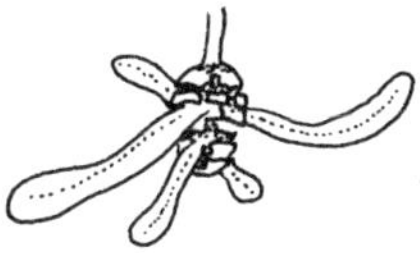

C
Taphrina occidentalis

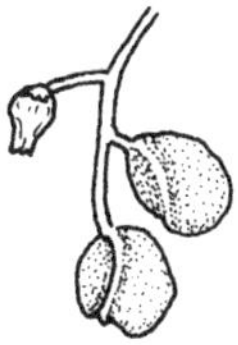

D
aphid gall

Group II—Leaf and Petiole Galls

Subgroup I—Bead and Pouch Galls

A. Bead galls up to 2 mm in diameter

B. Galls yellow-green, sometimes red, numerous, an alder leaves .. .. *Phytoptus laevis,* fig. 57E, p. 79

BB. Galls bright red to purple

C. Galls occurring single, many per leaf, on willows .... *Aculops tetanothrix,* fig. 57F, p. 155

CC. Galls single or combining to form coxcomblike clumps, pubescent, on poison oak .... *Aculops toxicophagus,* fig. 57G p. 166

AA. Bead or pouch galls larger than 2 mm in diameter

B. Galls club-shaped, elongate, yellow-beige, erect on wild cherries and plums .... *Phytoptus emarginatae,* fig. 58A, p. 179

BB. Galls globular, 2-4 mm, glossy, yellow-green, on walnut .... *Eriophyes brachytarsus,* fig. 58B, p. 144

E
Phytoptus laevis

F
Aculops tetanothrix

G
Aculops toxicophagus

Fig. 57

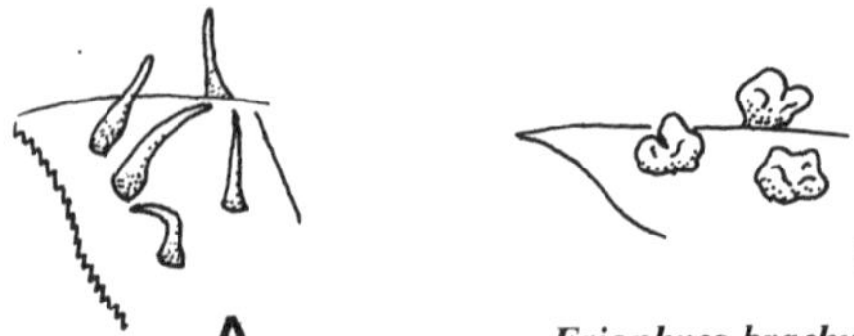

A *Phytoptus emarginatae*

B *Eriophyes brachytarsus*

Subgroup II—Erineum Galls

A. Erineum hairs red or rusty brown

B. Erineum magenta red, forming solid, feltlike mats on the upper surface of mountain maple ..... *Eriophyes calaceris,* fig. 58C, p. 92

BB. Erineum rusty-brown, on undersides in small pockets or covering large sections of the leaf, on oaks .... *Eriophyes mackiei,* fig. 58D, p. 105

AA. Erineum hair white, yellow, pink, or light brown

B. Erineum yellow, forming large pockets in between lateral veins, on undersides of walnut leaves .... *Eriophyes erineus,* fig. 58E, p. 144

BB. Erineum white, pink, or brown

C. Erineum white, undersides of big leaf maple .... *Eriophyes* sp., fig. 58E, p. 93

CC. Erineum white-yellow, pink, brown at maturity, in small round pockets on undersides, dorsal lumps light colored, on aspen .... *Phyllocoptes didelphis,* fig. 58F, p. 88

C *Eriophyes calaceris*

D *Eriophyes mackiei*

E *Eriophyes erineus*

F *Phyllocoptes didelphis*

Fig. 58

Subgroup III—Fold or Roll Galls

A. Fold galls along the margin or midrib vein

B. Fold gall bright red, along margins, oblong, 1 cm or more long, on manzanita .... *Tamalia coweni,* fig. 59A, p. 164

BB. Fold gall pink, red, or green, elliptical, fold toward underside of leaf, along margins and midrib in rows, on cottonwoods .... *Pemphigus populi-venae,* fig. 59B, p. 85

AA. Fold gall along the entire midrib vein or marginal roll galls turning inward, encompassing whole leaf, on spiraea .... *Rhabdophaga salicifolia,* fig. 59C, p. 175

A

Tamalia coweni

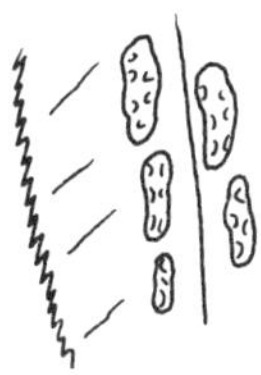

B

Pemphigus populi-venae

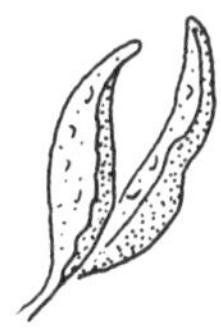

C

Rhabdophaga salicifolia

Subgroup IV—Mark Galls

A. Detachable from leaf without tearing leaf tissue

B. Galls with obvious bristles, hairs, or spines

C. Galls with soft, flexible, short or long hairs

D. Galls round, oblong, or club-shaped with short, 1 mm long or less hairs, on white oaks

E. Galls round-elliptic or oblong

F. Galls round, to 6 mm, with soft beige hairs, on undersides along midrib vein ...... *Andricus brunneus,* fig. 59D p. 115

Fig. 59

D

Andricus brunneus

FF. Galls elliptic, convoluted, rose color, clustered on midrib vein on underside .... *Andricus confertus,* fig. 61A, p. 120

EE. Galls club-shaped, covered with short beige pubescence, bulb near apex, on underside veins .... *Antron clavula,* fig. 61B, p. 124

DD. Galls with long, soft hairs

E. Galls on midrib vein, long, straw-yellow hairs, on white oaks .... *Andricus fullawayi,* fig. 61C, p. 117

EE. Galls with dingy-yellow to grey hairs on canyon live oaks .... *Andricus lasius,* fig. 61D, p. 137

CC. Galls with obvious stiff bristles, spines, or crystalline fibers

D. Galls with stout, stiff bristles, hairs, or spines

E. Galls round with spines or hairs projecting equally in all directions

F. Galls hard, blunt-tipped spines, red, lavender, or purple .... *Antron echinus,* fig. 61E, p. 122

FF. Galls with sharp, thin, red or brown spines or hairs

G. Galls succulent, red, sharp-tipped spines, on rose leaves .... *Diplolepis polita,* fig. 61F, p. 167

GG. Galls uniformly coated with stiff brown hairs, on white oaks .... *Sphaeroterus trimaculosum,* fig. 61G, p. 120

EE. Galls flat, disc- or funnel-shaped with spines projecting only laterally, on white oaks .... *Antron douglasii,* fig. 61H, p. 123

DD. Galls with brittle crystalline hairs or fibers, on white oaks

E. Galls large, forming caterpillarlike masses, with bent, thin, white, pink, rose, or brown fibers, which are long .... *Andricus crystallinus,* fig. 61I, p. 114

EE. Galls small, disc-shaped with central dome, crystalline fibers projecting laterally, on white oaks .... *Andricus stellaris,* fig. 61J, p. 118

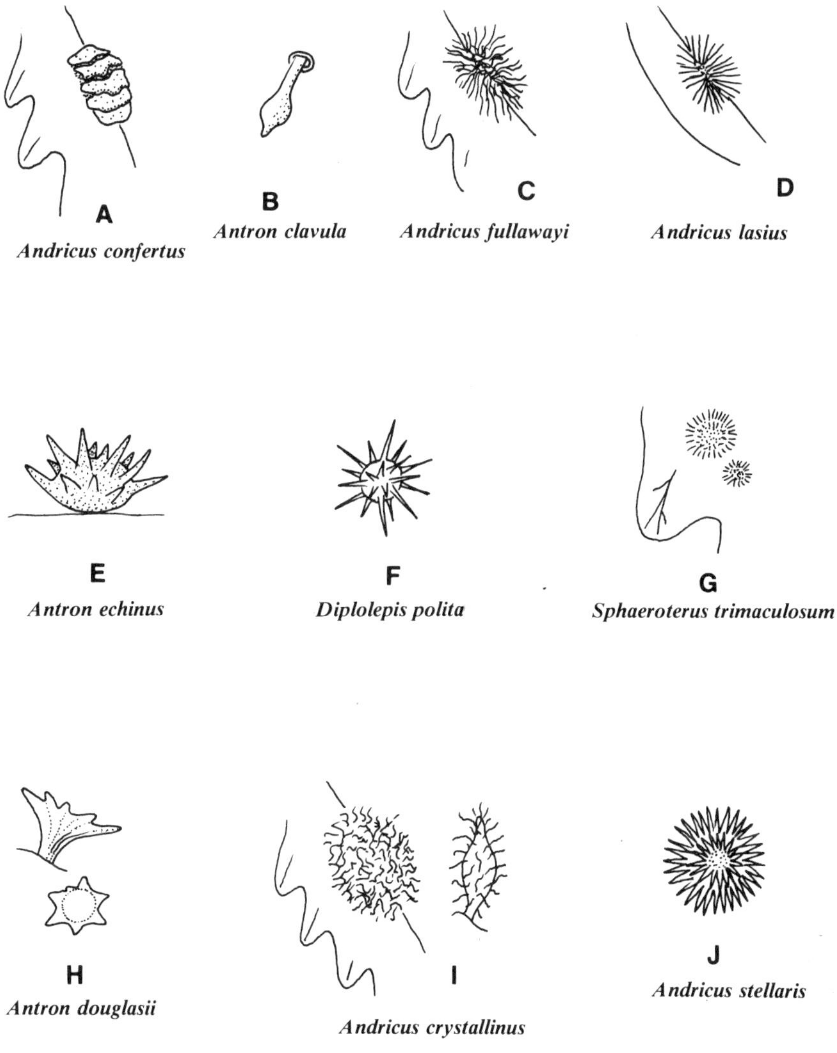
A
Andricus confertus
B
Antron clavula
C
Andricus fullawayi
D
Andricus lasius
E
Antron echinus
F
Diplolepis polita
G
Sphaeroterus trimaculosum
H
Antron douglasii
I
Andricus crystallinus
J
Andricus stellaris

Fig. 61

BB. Galls without obvious hairs, spines or crystalline fibers

C. Galls flat, disc- or cup-shaped, with or without a stalk

D. Galls flat or disc-shaped, without upturned edges, up to 1 cm wide

E. Galls normally 1 cm wide with margins touching the leaf surface, convex or concave center

F. Galls smooth edged, convex center, beige, on canyon live oaks .... *Paracraspis guadaloupensis,* fig. 62A, p. 138

FF. Galls smooth to crenate at the edge, often with depressed center, yellow-green, on white oaks .... *Andricus pattersonae,* fig. 62B, p. 119

EE. Galls 5 mm or less, margins do not touch leaf, connected centrally, flat, radiating colors, glossy, on white oaks .... *Andricus parmula,* fig. 62C, p. 118

DD. Galls cup-shaped with upturned edges, with or without a stalk

E. Galls to about 5 mm wide, crenate margin, red, brown, yellow, on white oaks ...... *Andricus gigas,* fig. 62E, p. 117

EE. Galls to 1 cm or more, often rising on a stalk

F. Galls stalked, prominent cup or depression on top, on canyon live oaks .... *Paracraspis patelloides,* fig. 62D p. 138

FF. Galls stalked, cup-shaped when young, top flattens in age, appearing like mushroom, spring gall on black oaks .... *Callirhytis pomiformis,* fig. 62F, p. 100

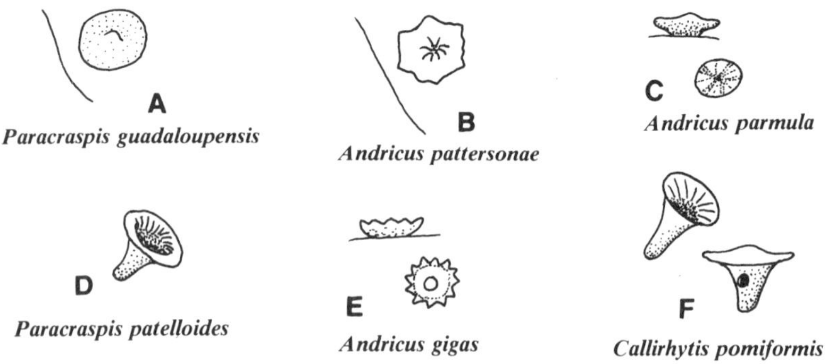

Fig. 62

CC. Galls round, cone-shaped, laterally compressed, or with a thin, hairlike stalk

D. Galls round or laterally compressed, small

E. Galls round, under 2 mm wide

F. Galls yellow-green, on tops of leaves of black oaks .... *Dryocosmus minusculus,* fig. 63A, p. 103

FF. Galls about 1 mm, chestnut brown, on lower surface of white oaks .... *Neuroteras saltatorius,* fig. 63B, p. 120

EE. Galls green-tan, laterally compressed with a point at each end, on veins, undersides of black oaks .... *Dryocosmus dubiosus,* fig. 63C, p. 96

DD. Galls cone-shaped, pyramidal, or thinly stalked

E. Galls cone- or pyramidal-shaped, to 6 mm tall

F. Galls solid red with basal edges flaring out and up, centrally attached, on white oaks .... *Andricus kingi,* fig. 63D, p. 116

FF. Galls pyramidal, white to cream with red vertical stripes, basal edges touching leaf, on white oaks .... *Andricus atrimentus,* fig. 63E, p. 125

EE. Galls with hairlike, thin stalk, bulge near pointed apex, to 2 cm long, on white oaks .... *Dros pedicellatum,* fig. 63F p. 126

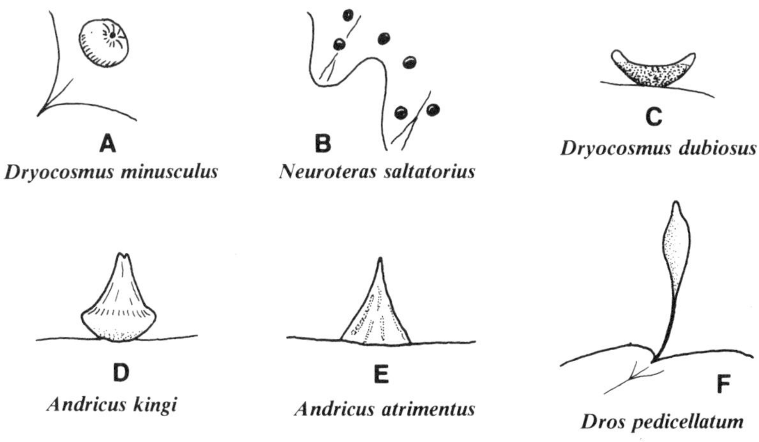

Fig. 63

AA. Galls not detachable without tearing leaf tissue

B. Galls round, elongate, or conical-tubular, with or without hairs

C. Galls round to elongate

D. Galls round, one or more per leaf

E. Galls smooth, brown, one per leaf, on midrib vein, dorsal side of black oaks *Callirhytis perfoveata,* fig. 64A, p. 104

EE. Galls red, green, one or more per leaf, with small tubercles, along margins and midrib vein, on willows .... *Pontania pacifica,* fig. 64B, p. 153

DD. Galls elongate, brown, smooth, encompassing petiole and part of midrib vein of the leaf, on black oaks .... *Callirhytis flora,* fig. 64C, p. 103

CC. Galls conical, tubular, or thorn-shaped

D. Galls red, yellow, without noticeable hairs

E. Volcano shape with depression at the tips, on undersides of leaves of willows .... *"Cecidomyia" salicisverruca,* fig. 64D, p. 154

EE. Thorn-shaped, fresh in spring, on undersides of leaves of canyon live oak .... *Heteroecus pacificus,* fig. 65C p. 130

DD. Galls grey-green, pubescent, along margins of leaves, on petioles and midrib, depression at apex, on sage .... *Rhopalomyia salviae,* fig. 64E, p. 171

BB. Gall encompassing whole or part of leaf, causing distortion, swelling, and yellowing, leaf curl on cherries and plums .... *Taphrina flectans,* fig. 64F, p. 180

A
Callirhytis perfoveata

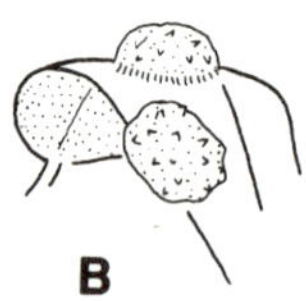

B
Pontania pacifica

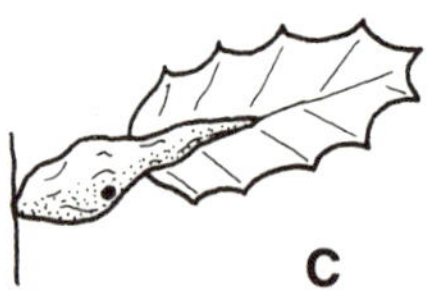

C
Callirhytis flora

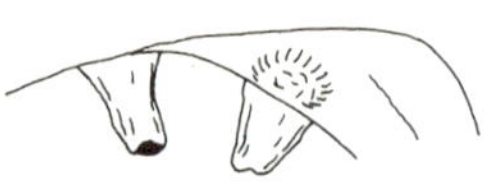

D
"Cecidomyia" salicisverruca

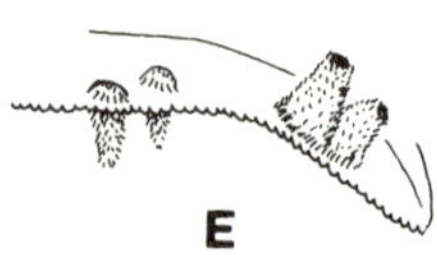

E
Rhopalomyia salviae

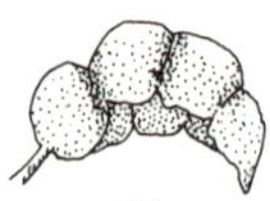

F
Taphrina flectans

Fig. 64

Subgroup V—Petiole Galls

A. Galls soft and fleshy with noticeable slit along one side, red-green, hollow

B. Galls globular, at the base of the leaf, often overlapping into leaf tissue, single or in pairs, slit with spiral twist, on cottonwoods .... *Pemphigus populicaulis,* fig. 65A, p. 84

BB. Galls with transverse, horizontal or sometimes spiral twisting slit, often with prominent lips along slit, causing petiole to bend, on cottonwoods ...... *Pemphigus populitransversus,* fig. 65B, p. 82

AA. Galls solid, hard, brown with numerous emergence holes, also often encompassing leaf midrib, on black oaks .... *Callirhytis flora,* fig. 64C, p. 103

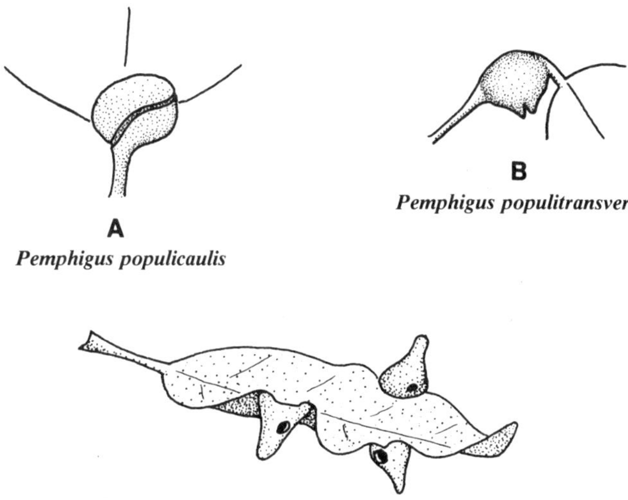

A *Pemphigus populicaulis*

B *Pemphigus populitransversus*

C. Bisexual generation galls of *Heteroecus pacificus*

Fig. 65

Group III—Stem Galls[1]

Subgroup I—Bud Galls

A. Galls with scales, bracts, leaflike projections or round, oblong or fleshy

B. Galls with scales, bracts, or leaflike projections

C. Galls with bud scales more than 3 mm wide, on conifers

D. Galls and scales elongate, sharp pointed, in four opposing rows, to 3 cm long, on cypress .... *Walshomyia cupressi,* fig. 66A, p. 80

DD. Galls globular, 3 cm wide, scales broad to 1 cm, green to brown, resembles small cone or burr, on junipers .... *Oligotrophus betheli,* fig. 66B, p. 89

CC. Galls round to cone-shaped with thin, broad leaflike projections or linear filaments in rosettes

D. Galls round, large, to 4 cm in diameter, with thin pointed bracts in numerous rosette patterns, brittle when brown, on creosote bush .... *Asphondylia auripila,* fig. 66C, p. 161

DD. Gall short to long pyramidal cone with distinct leaflike bracts radially arranged, on willows .... *Rhabdophaga brassicoides,* fig. 66D, p. 150

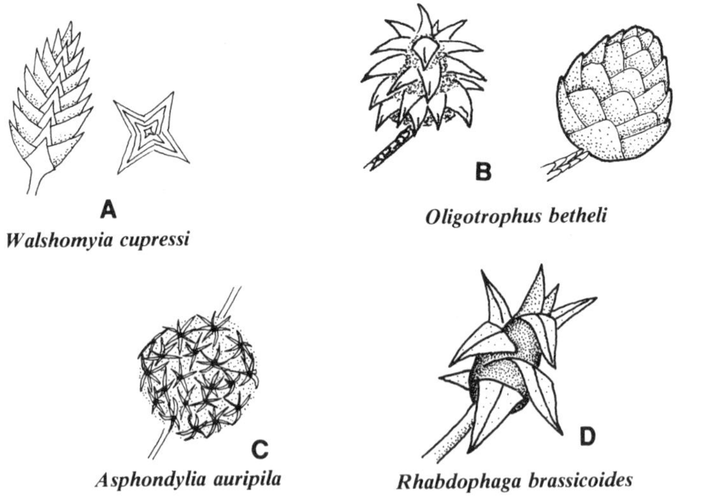

Fig. 66

1. Some overlap may occur between subgroups I, II, and III.

BB. Galls without scales (except one species) or leaflike projections, but fleshy or hard, smooth and hairless or covered with short or long hairs

C. Galls smooth and hairless or if hairs present, they are short

D. Galls hard, round to urn-shaped

E. Galls round, hard, to 1 cm, green, mottled when fresh, brown by autumn, short fuzzy pubescence when fresh in spring, at the tips of shoots, on black oaks .... *Callirhytis agrifoliae,* fig. 67A, p. 102

EE. Galls urn-shaped, berrylike, to 1 cm high, often with grey bloom, clasping base, young galls closed at apex, mature galls split at apex into 3-5 lobes, on junipers .... *Walshomyia sabinae,* fig. 67B, p. 89

DD. Galls spongy, fleshy, with or without hairs and leaves or leaflets that may be stunted

E. Galls without obvious hairs, fleshy and smooth, often with small leaves

F. Galls globular, knobby, to 2 cm, green to purple-red, at the tips of shoots on coyote brush .... *Rhopalomyia californica,* fig. 67C, p. 157

A

Callirhytis agrifoliae

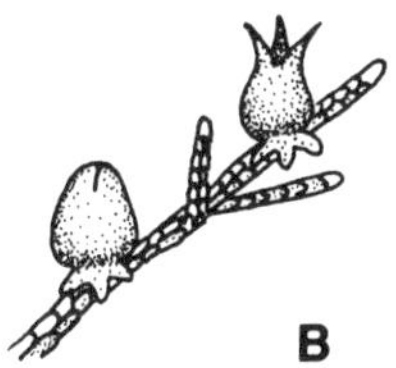

B

Walshomyia sabinae

C

Rhopalomyia californica

Fig. 67

FF. Galls rounded but marked with overlapping bud scales much expanded and terminating in curved tips, without stunted leaves, on tips of shoots of silk tassel .... *Asphondylia garryae,* fig. 68A, p. 174

EE. Galls to 3 cm wide, covered with dense, short, bristly hairs, overlapping bud scales, dwarfed leaflets, on stems of lupine .... *Dasineura lupini,* fig. 68B, p. 162

CC. Galls round to oblong, covered with dense mat of long hairs

D. Galls under 1 cm, with loose, long, white cottonlike hairs, at base of leaves of sagebrush .. *Rhopalomyia floccosa,* fig. 68C p. 173

DD. Galls well over 1 cm wide, long hairs present

E. Galls to 3 cm round, with dense mat of thin, white hairs, encircling stems of saltbush .... *Asphondylia atriplicis,* fig. 68D, p. 170

EE. Galls with white, green-red tinged hairy, pointed lobes, overall color white-green, on stems of sagebrush .... *Rhopalomyia* sp., fig. 68E, p. 173

AA. Galls irregular, knobby, elongate, often continuous masses to 5 cm long, with fine short hairs, causing stem distortion on aspen .... *Eriophyes parapopuli,* fig. 68F, p. 86

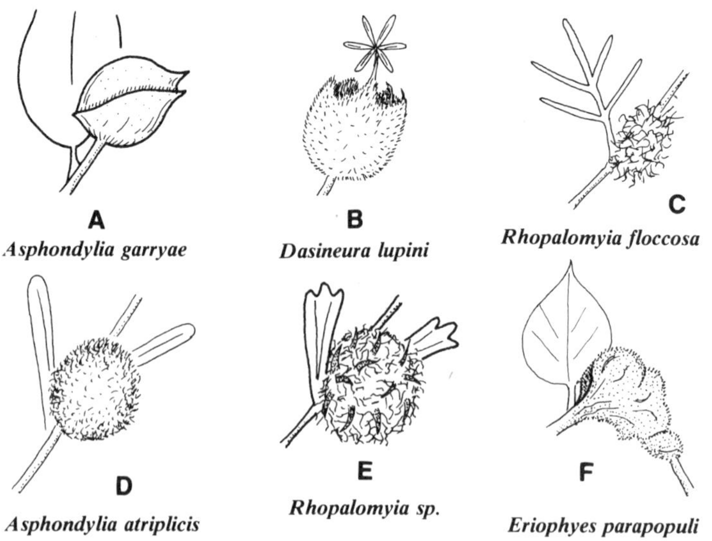

A *Asphondylia garryae* B *Dasineura lupini* C *Rhopalomyia floccosa*

D *Asphondylia atriplicis* E *Rhopalomyia sp.* F *Eriophyes parapopuli*

Fig. 68

Subgroup II—Detachable Galls

A. Galls round to oblong-ovoid

B. Galls normally 3 cm or more in diameter or length, without leafy lobes or thin spines

C. Galls with occasional wartlike knobs, or short, thick spines or protuberances

D. Galls to 10 cm long, potato-shaped to round, smooth, but with dull, wartlike knobs, on white oaks .... *Andricus californicus,* fig. 69A, p. 107

DD. Galls to about 4 cm wide, round, often with numerous short, thick, sharp-pointed surface projections, on black oaks .... *Callirhytis pomiformis,* fig. 69B, p. 100

CC. Galls to 6 cm, oblong, smooth surface, thin outer skin cracks and peels, brown, on canyon live oaks .... *Disholcaspis truckeensis,* fig. 69C, p. 134

BB. Galls generally under 3 cm

C. Galls 1-3 cm in diameter or length

D. Galls round, solid, smooth or with leaflike lobes

E. Galls with smooth, silver-grey waxy coating, hard, 1-2 cm wide, on white oaks ..... *Disholcaspis canescens,* fig. 69D p. 108

A

Andricus californicus

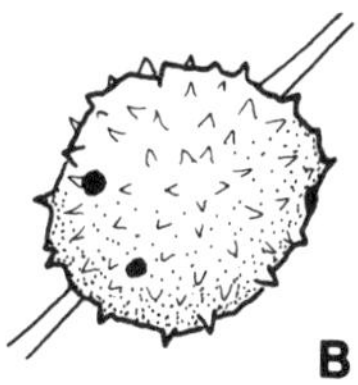

B

Callirhytis pomiformis

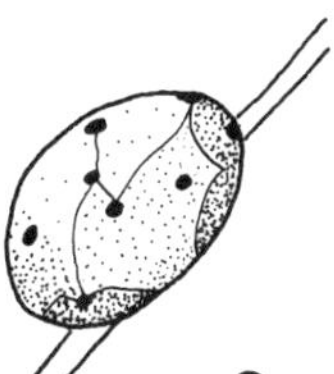

C

Disholcaspis truckeensis

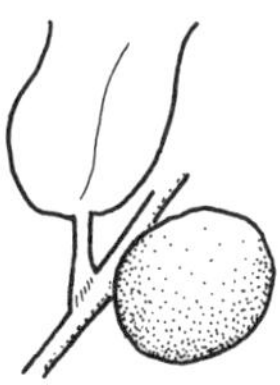

D

Disholcaspis canescens

Fig. 69

EE. Galls covered with numerous leaflike lobes, green-red, brown by autumn, 1-3 cm in diameter, but occasionally to 5 cm, slightly spiny, on rose .... *Diplolepis californica,* fig. 71A, p. 168

DD. Galls thin and hollow or oblong and solid

E. Galls round, smooth with shallow surface depressions, hollow, thin skinned, filled with radiating fibers leading to central chamber, on huckleberry oaks .... *Trichoteras vaccinifoliae,* fig. 71B, p. 136

EE. Galls solid, mostly smooth, kidney-shaped, mottled, on canyon live oaks *Andricus reniformis,* fig. 71C, p. 135

CC. Galls normally 1 cm or less, on white oaks

D. Galls round, relatively smooth, without projections, or round-sided, flat-topped

E. Galls round, solid, smooth to wrinkled surface, single or in groups

F. Galls without a neck, wrinkled, often difficult to distinguish from next choice .... *Besbicus conspicuus,* fig. 71D, p. 110

FF. Galls often with a neck or stalk attachment along coastal areas, similar to preceding .... *Disholcaspis washingtonensis,* fig. 71E, p. 110

EE. Galls round, but with flat or convex, pitted top, light color on sides, dark on top, attracts ants .... *Disholcaspis eldoradensis,* fig. 71F, p. 110

DD. Galls round with a single or multiple sharp or obtuse projections

E. Galls red, yellow, orange, with numerous blunt-tipped projections .... *Disholcaspis corallina,* fig. 71G, p. 109

EE. Galls round, smooth, red and yellow mottled or solid lemon yellow, with a single, long, sharp-pointed projection or nipple .... *Disholcaspis plumbella,* fig. 71H, p. 112

A

Diplolepis californica

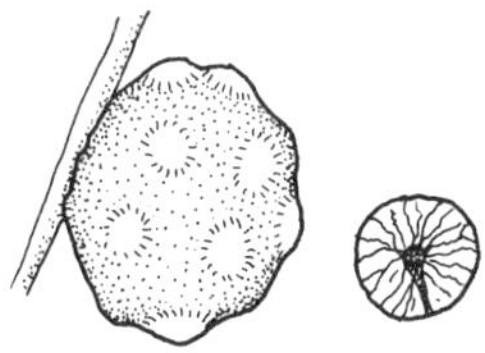

B

Trichoteras vaccinifoliae

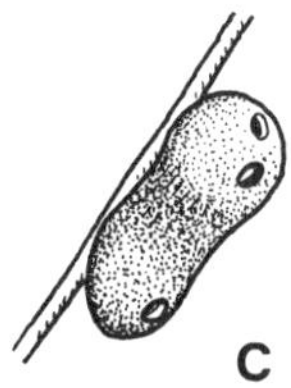

C

Andricus reniformis

D

Besbicus conspicuus

E

Disholcaspis washingtonensis

F

Disholcaspis eldoradensis

G

Disholcaspis corallina

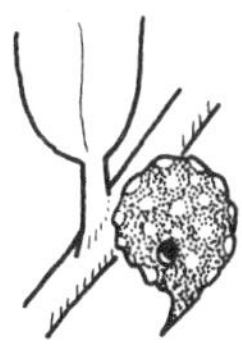

H

Disholcaspis plumbella

Fig. 71

AA. Galls cylindrical, linear, elongate or other shape, if globular, then covered with yellow-beige hairs

B. Galls elliptic-linear, with or without yellow hairs, on canyon live oak

C. Galls masked with yellow hairs

D. Long yellow hairs rub off easily, exposing smooth, elliptic gall beneath, gall tapered at each end with central bulge .... *Heteroecus dasydactyli,* fig. 72A, p. 128

DD. Short, dense mat of dingy, yellow hairs which do not rub off easily, gall normally not radically tapered at each end, nearly uniform in dimensions .... *Heteroecus melanoderma,* fig. 72B, p. 128

CC. Galls without hairs, elliptic, to 3 cm long, smooth, mottled red and green, or solid color, tapered at each end, some forms pear-shaped with curved nipple .... *Heteroecus pacificus,* fig. 72C, p. 130

BB. Galls other shapes

C. Galls occurring in clusters, rarely isolated

D. Galls appear mushroom-shaped with broadly clasping bases, smooth sides, pitted cap, attracts ants, on white oaks .... *Disholcaspis prehensa,* fig. 73A, p. 112

DD. Galls closely clustered in rows with crested, dark brown, cracked ridge on top, smooth, light-colored sides, on canyon live oaks ...... *Disholcaspis chrysolepidis,* fig. 73B, p. 132

CC. Galls generally isolated, appearing in two distinct segments, constricted in the middle, round base, pointed, conelike cap, on canyon live oaks ..... *Heteroecus sanctae-clarae,* fig. 73C, p. 132

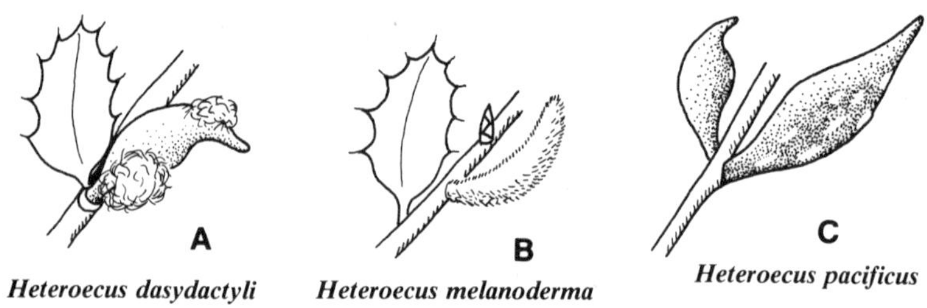

A — ***Heteroecus dasydactyli***

B — ***Heteroecus melanoderma***

C — ***Heteroecus pacificus***

Fig. 72

A
Disholcaspis prehensa

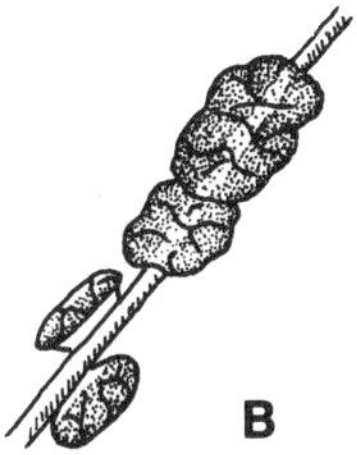

B
Disholcaspis chrysolepidis

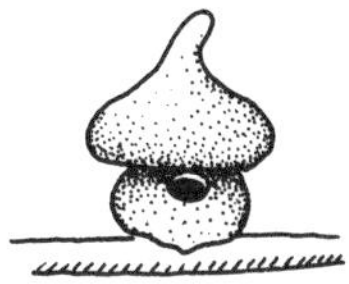

C
Heteroecus sanctae-clarae

Subgroup III—Integral Galls

A. Galls elliptic, tapered at one or both ends, but not radically, abrupt swellings

B. Galls naked, without buds, shoots or leaves protruding, to 8 cm long, on trees

C. Galls rough, deeply fissured, cracked or peeling

D. Galls bark color with several short, jagged, narrow slits or cracks, through which the larval chambers are expelled, on black oaks .... *Callirhytis perdens,* fig. 73D, p. 100

DD. Galls appear broken, with large segments of the bark peeled back or absent revealing the nest of larval chambers, on canyon live oaks ..... *Dryocosmus asymmetricus,* fig. 73E, p. 127

CC. Galls smooth to slightly knobby, without deep fissures

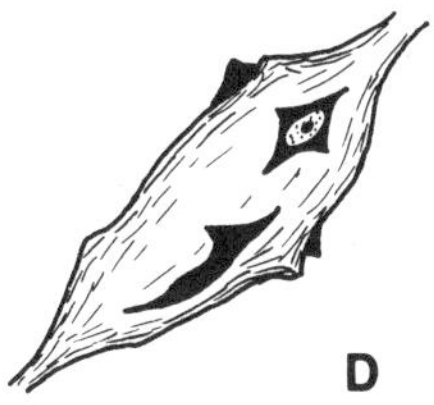

D
Callirhytis perdens

E
Dryocosmus asymmetricus

Fig. 73

D. Galls smooth, robust swellings, small emergence holes, on canyon live oaks .... *Andricus spectabilis,* fig. 74A, p. 134

DD. Galls more cylindrical, smooth or knobby, on white oaks .... *Andricus chrysolepidicola,* fig. 74B, p. 106

BB. Galls with reduced leaves, needles, buds, or shoots, on trees or shrubs, some with deep, orange-colored fissures

C. Galls with pronounced shoots emanating from gall

D. Galls with dense, yellow, orange, or green fleshy shoots, often with small flowers or seeds, on pines .... *Arceuthobium campylopodum, A. americanum,* fig. 74C p. 141

DD. Galls with green to bark-colored leafy shoots, witches' brooms, fissures orange or white-colored, on coyote brush, *Baccharis* sp. .... *Puccinia evadens,* fig. 74D, p. 159

CC. Galls with reduced or normal leaves or needles

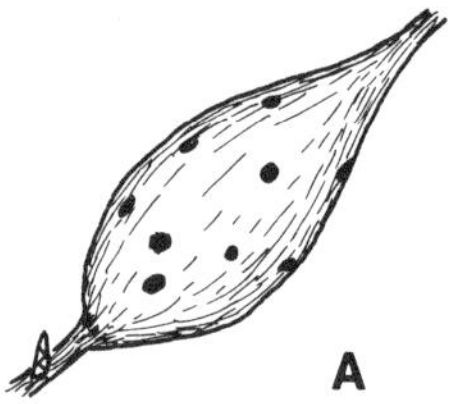

Andricus spectabilis

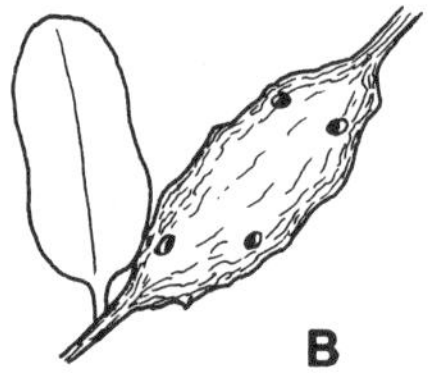

Andricus chrysolepidicola

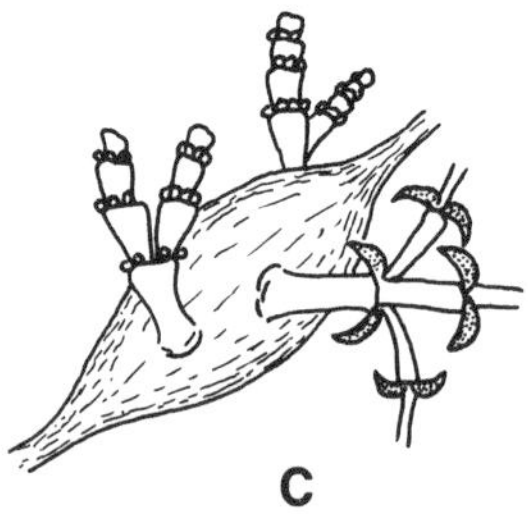

Arceuthobium campylopodum, A. americanum

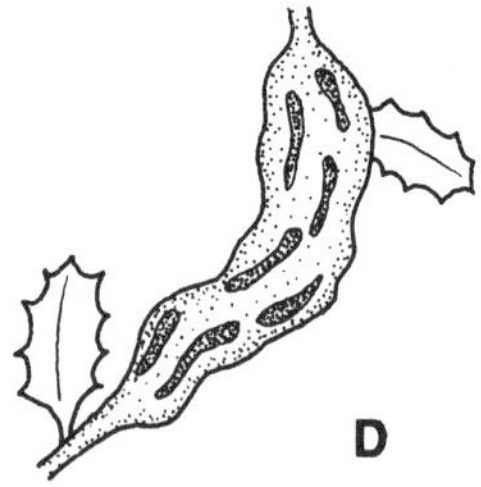

Puccinia evadens

Fig. 74

D. Galls with normal needles surrounding the gall, with opening or slit at the base of each needle at maturity, on spruce .... *Adelges cooleyi,* fig. 75A, p. 142

DD. Galls with or without leaves, leaflets, or buds

E. Galls with or without leaves or leaflets

F. Galls elliptic, tapered equally at both ends, with or without subnormal leaves, gall smooth, hairless, on coyote brush .... *Gnorimoschema baccharisella,* fig. 75B p. 158

FF. Galls linear, pubescent, with stunted leaflets, on lupine .... *Neolasioptera lupini*, fig. 75C, p. 163

EE. Galls glossy, with scattered normal or aborted buds, red-purple to yellow-green, on willows .... *Euura scoulerianae,* fig. 75D, p. 148
...... *E. breweriae, E. lasiolepis, E. exiguae,* fig. 75E, p. 146

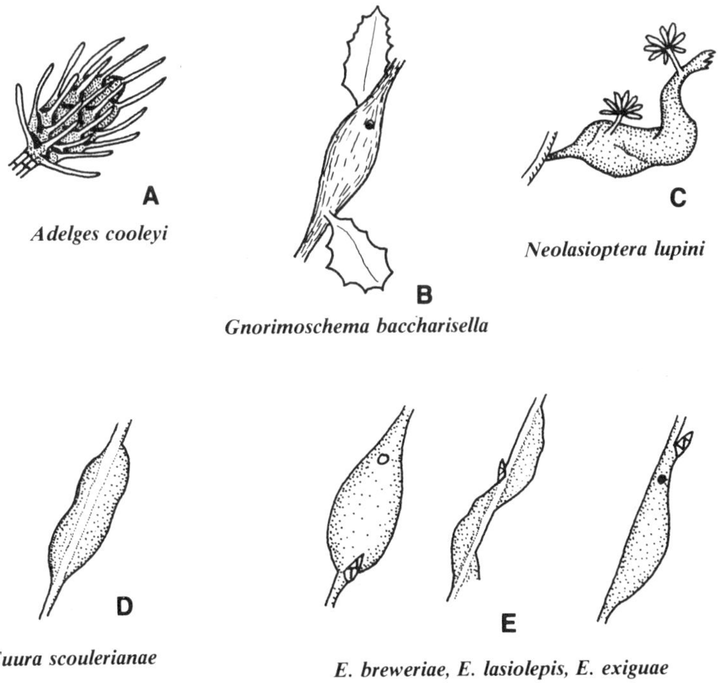

Fig. 75

AA. Galls abrupt swellings, often lumpy, fissured, or radically bent or twisted

B. Galls under 2 cm in diameter, short, knobby, globular-potato-shaped, glossy, not pure black, on willows ....
.... *Euura geyerianae,* fig. 77A, p. 147
.... *E. lemmoniae,* fig. 77B, p. 149
.... *Rhabdophaga salicisbatatus,* fig. 77C, p. 152

BB. Galls over 2 cm in diameter, if less, then they are black

C. Galls smooth, finely cracked to knobby

D. Galls smooth, grey, finely cracked, to 10 cm long by 4 cm wide, globular, on black oaks ....
.... *Callirhytis quercussuttoni,* fig. 77D, p. 98

DD. Galls 4-8 cm long, green to light brown, slightly knobby, often bent, with sparse, fine, bristly hairs, on thimbleberry ....
.... *Diastrophus kincaidii,* fig. 77E, p. 177

CC. Galls deeply fissured or minutely pitted, round or linear and bent

D. Galls round, deeply fissured, orange spores in season, to 30 cm wide, on pines .... *Peridermium harknessii,* fig. 77F p. 140

DD. Galls black, bent, twisted, elongate to 10 cm, with erupting edges, smooth in general, but with minute pits over the surface, on one side of stem, on cherries and Sierra plum
.... *Dibotryon morbosum,* fig. 77G, p. 181

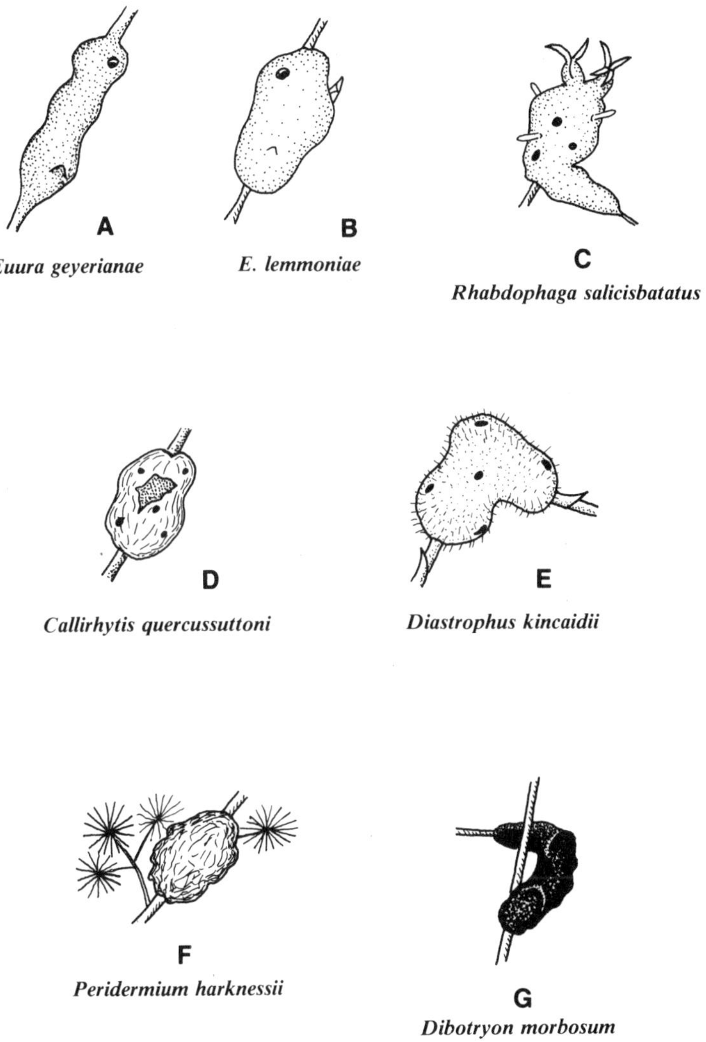
A
Euura geyerianae
B
E. lemmoniae
C
Rhabdophaga salicisbatatus
D
Callirhytis quercussuttoni
E
Diastrophus kincaidii
F
Peridermium harknessii
G
Dibotryon morbosum

Fig. 77

4

DESCRIPTIONS OF GALL SPECIES

THE detailed descriptions of the individual galls included in this guide are grouped according to the host plants and not the generic groups or families of gall organisms. The host plants are arranged alphabetically by common name. They proceed from trees to shrubs.

Alder Galls

There are four common alder trees that occur along the stream banks of California and extend north into British Columbia and Alaska. These include: the sitka alder, *Alnus sinuata;* the red alder, *A. oregona;* the white alder, *A. rhombifolia;* and mountain alder, *A. tenuifolia.* Both the sitka and mountain alder appear as small trees or shrubs. On the other hand, white and red alders attain stately heights of 21 m at maturity. Sitka alder is found mainly in the extreme northern counties of California. Mountain alder occurs in the Sierra Nevada from Tulare County north. Red alder grows at elevations generally under 300 m from Santa Cruz County north, while white alder is found up to 450 m in the Sierra and the central and southern coastal ranges.

There are two common and abundant alder galls that, strangely enough, appear on all four species of alders mentioned. One gall is induced by a fungus on the female cones of the trees, while another gall is induced by mites on the leaves.

The alder tongue gall fungus, *Taphrina occidentalis,* induces broad, flat, twisting bract galls on the cones in the spring and early summer months. Actually, the galls are cone bracts or scales that expand wildly into tongue-shaped enlargements. These **tongue** galls grow to 4 cm long by 5 mm wide. The galls are

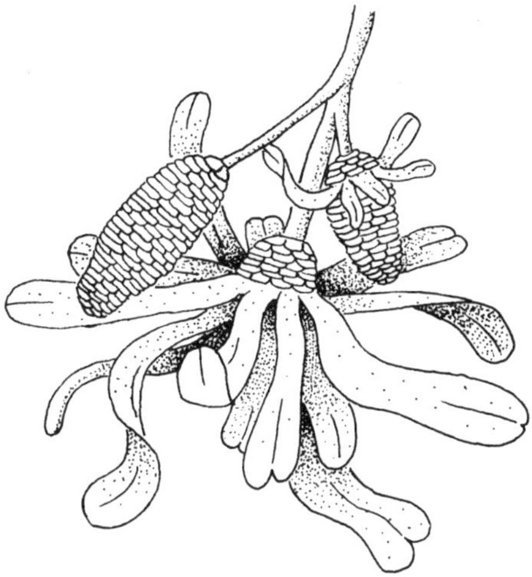

Fig. 79. Alder tongue gall, *Taphrina occidentalis*

usually green with red along the margins. Some cones are so thoroughly infested with this fungus that they are completely obscured. Old cones remain on the trees for a year or so after the initial formation of the galls.

Tongue galls on alders in the Sacramento Valley mature earlier than those of the Sierra due to the difference in timing of cone development. Usually by August tongue galls in the Sierra begin to brown out and wither. The spores of this fungus survive the winter and reinfect the tree with the coming of spring. Actually, this fungus resembles a yeast during the asexual part of its life during the winter. As the cones develop and near their full size, the serpentine enlargements of the bracts form. This fungus occurs throughout the western United States.

The alder bead gall mite, *Phytoptus laevis,* induces round, yellow to green bead galls on the upper surfaces of the leaves of alders. These small, 2 mm in diameter bead galls reveal their rough surface under close inspection with a hand lens. As with other bead galls induced by mites, there is an opening on the underside. The galls appear singly or in large numbers per leaf.

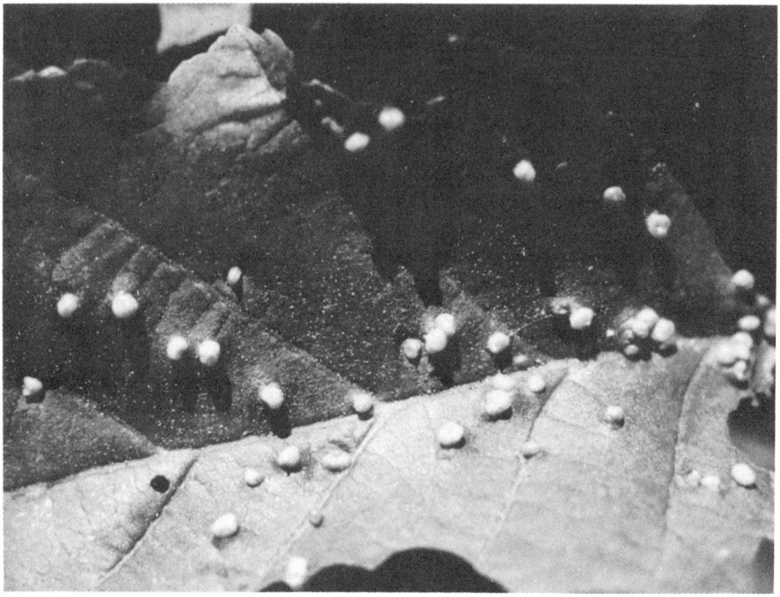

Fig. 80. Alder bead gall, *Phytoptus laevis*

They rarely become noticeable until after the leaves have reached their full development. The adult mites that emerge from these summer galls probably spend the winter in the fissures of the tree bark. No apparent damage is done to the leaves. This mite was first described in Europe and is widespread in the western states.

Cypress Galls

Within the stunted, bonsai-like forests of pygmy cypress, *Cupressus pygmaea,* as well as on isolated individuals, in Mendocino County an interesting well-camouflaged midge gall appears on this host. The lofty sargent cypress, *C. sargentii,* of Sonoma County also hosts this gall midge.

The cypress gall midge, *Walshomyia cupressi,* induces large, monothalamous, four-sided bud galls in the summer. Architecturally, the galls are fascinating in that they are massive

Fig. 81. Cypress gall, *Walshomyia cupressi*

replicas of the buds with four series of elongated, pointed, overlapping scales. At maturity the galls measure up to 3 cm long by 1 cm wide. Each gall starts as a swollen, beige, globular bud. But as development proceeds, the overlapping scales separate from each other near their tips. At this stage, looking down the axis of the gall reveals the four-sided pattern. In September young galls 5 mm long can be found growing. These are beige-yellow. At the same time, however, one can find the larger, mature green galls. Abandoned galls do not remain on the tree for long. The galls normally form in large terminal clusters. The adults emerge through the side near the apex after nearly two years in the larval stage within the galls. Adults live only for a few days in the spring while they reproduce and oviposit. Vigorously growing cypress trees seem to host more of these midges than the more stunted specimens growing nearby.

Cottonwood and Aspen Galls

Along the streams, rivers, and high-mountain meadows grows a specialized group of deciduous trees known as the poplars. The poplar group includes the cottonwood and aspen trees. Three of the four native species in California are extremely common and widespread. These include quaking aspen, *Populus tremuloides;* black cottonwood, *P. trichocarpa;* and Frémont cottonwood, *P. fremontii.* Quaking aspen dominates high-mountain meadows from San Bernardino County north through the Sierra to Alaska. Black cottonwood is found along streams at high elevations from San Diego County north to Alaska. Frémont cottonwood occurs along low-elevation stream and river banks in the valleys, coastal areas, and south into the desert and Nevada. The fourth native poplar, *P. angustifolia,* is found in scattered locations in Inyo County. Three species of poplars that also host gall organisms have been introduced into California. Lombardy poplar, *P. nigra* var. *italica, P. balsamifera,* and the silver poplar, *P. alba,* host unique galls; but the emphasis here will lie with indigenous trees.

These trees are attacked and galled by a variety of organisms, including fungi, eriophyid mites, and aphids. The fungus cankers commonly found on aspen and Frémont cottonwood are difficult to isolate and identify. Aphids are by far the most prominent gall-inducing organisms on cottonwoods and aspens in this region. Most of the aphid galls are formed on the petioles of the leaves, while the mite galls are found on the buds and leaves.

As mentioned earlier, the taxonomy of gall aphids needs serious clarification. The names used here are the best available and may be changed with continuing research.

The gall aphid *Pemphigus populitransversus* (?) induces fleshy, orbicular, monothalamous galls on the petioles at the base of the leaves of Frémont cottonwood. The galls measure up to 14 mm in diameter. They are characterized by a prominent transverse or angular slit. The edges of the slits often protrude slightly, creating the appearance of lips. A tiny opening is situated in the center of

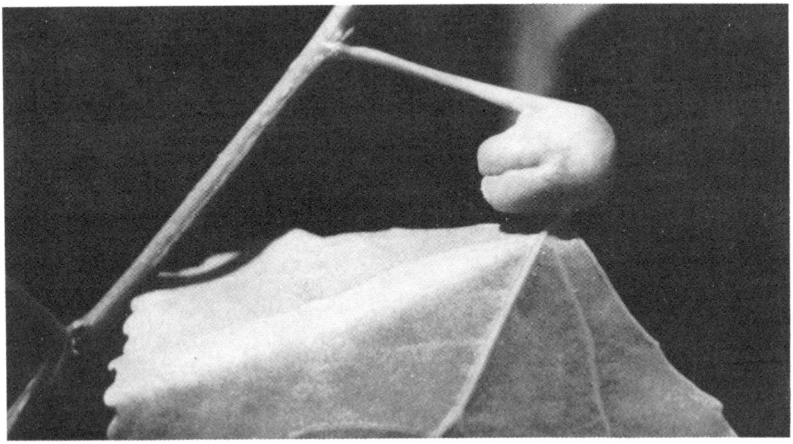

Fig. 83. Aphid gall, *Pemphigus populitransversus*

each slit. The galls are smooth, green-red, and located to one side of the petioles. Frequently the rapid growth of the galls forces the petioles to bend radically to a 90-degree angle just above the galls.

The swelling begins its development around the stem mother as she feeds, providing the stimulus for gall formation. Eventually, the gall tissue encloses the female completely. The female lays eggs and the gall rapidly fills with hungry offspring. The white, waxy secretions of the young often conceal them. As the gall becomes overcrowded, the young exit and crawl about on the leaves and stems of the host plant. Aphids are present in the galls all summer, even though there appears a constant emergence during the entire season. In the fall, as the leaves change color, wither, and drop to the ground, the galls normally remain green or red and succulent for up to two weeks. As the galls begin to dry up, the transverse slits separate, allowing the escape of the larger aphids.

Along the Russian River in Sonoma County, galls on Frémont cottonwood develop at varying times depending on the height of winter flooding. Those parts of the trees above the high-water mark produce galls earlier than the lower sections or younger individual trees. Winter flooding may destroy the overwintering aphids from the previous season's galls. Thus, galls on lower sections do not develop until aphids have moved down from higher, already formed galls and begun feeding. Cottonwoods in the interior warm valleys develop galls earlier than do trees of the cooler coastal valleys. Although this aphid has been reported on quaking aspen in Canada, I have never found it on anything but Frémont cottonwood. Lacewing insect larvas and certain birds are known to prey on the aphids of these galls. Of greater consequence to the aphids is a fungus that develops within the galls, which shortens the duration of reproduction and causes high mortality in the early nymphal stages.

The gall aphid *Pemphigus populicaulis* (?) induces globular, fleshy, monothalamous, succulent galls on the petioles at the base of the leaves of black cottonwood in the Sierra Nevada. The galls, which measure up to 15 mm in diameter, almost always bear spiral slits running at an angle away from the petioles. These galls normally develop in such a way that they usually encompass part of the lower section of the leaves. The slits normally occur on the same side of the gall as the underside of the leaf. These lopsided galls protrude from the bottom surface of the leaf and petiole.

Fig. 84. Two aphid galls, *Pemphigus populicaulis,* on the same petiole

They appear to be formed through massive swelling and twisting. They occur singly, but very often one finds two galls side by side on the same petiole. Gall formation begins in the late spring. Galls remain green and active until September and October. The differences between this gall and others caused by *Pemphigus* spp. may be attributed to the type of cell tissue selected by the stem mother for feeding and / or secretions injected into the plant. This aphid is known to use water parsley, *Oenanthe sarmentosa,* as a secondary host, but it has not yet been determined whether the aphid uses the same plant in California.

In addition to the predators already mentioned, ants have been observed chewing through the walls of the galls. On several occasions I have found ants in the galls without a single aphid present. This raises questions that still remain unanswered. Did the aphids leave upon the arrival of the ants? Or were the aphids and their sweet exudate consumed by the ants? Perhaps, under the stress of hunger, ants cannot pass up any nutritional opportunity.

The aphid *Pemphigus populi-venae* (?) induces elliptical, wrinkled, monothalamous fold galls on the leaves of black cottonwood. The pale green, yellow-red galls measure 10 mm long by 4 mm wide. They usually are found lined up in rows on both sides of the midrib vein or along the leaf margins. The galls are most noticeable from the dorsal surface of the leaves. The aphids feed on the underside of the leaves, which causes the leaf tissue to swell and bulge upward and around the aphids. A linear slit on the underside of the leaf is the entrance to the bulging gall found on the dorsal side. Under heavy galling, an individual leaf may be drastically distorted.

A few cottonwoods growing near each other may support thousands of these or other *Pemphigus* galls, while downstream the cottonwoods are barren of aphid galls of the same species. There is some evidence that individual clones of trees are immune or resistant to the aphid attacks, while other clones are very susceptible. Although numerous questions concerning the chemistry of this intriguing relationship are raised, answers are

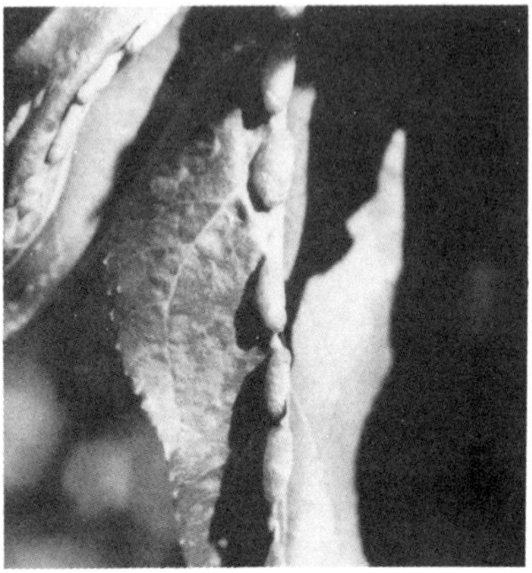

A. Aphid galls, *Pemphigus populi-venae,* developed along the midrib vein. The slits through which the aphids emerge are located on the underside of the leaves.

B. The galls of *Pemphigus populi-venae* also form along each margin of the affected leaf.

Fig. 86

not yet clear. This aphid is known to use sugar beets and plants of the family Chenopodiaceae as alternate hosts.

In the Central Valley near Davis, California, as well as throughout the United States, large populations of the aphid *Pemphigus bursarius* induce galls on Lombardy poplar. Known as the European lettuce aphid, this species is a pest on lettuce, which is its alternate host. Of particular interest with respect to these aphids is the incredible reproduction capacity of the stem mothers. As many as 1,091 offspring emerged from a single *Pemphigus nortoni* gall on Frémont cottonwood near Davis during a study by Lange and Grigarick.

The poplar bud gall mite, *Eriophyes parapopuli,* induces nodular, rough, minutely pubescent bud galls on quaking aspen. The galls often involve large sections of the stems near the buds. This makes them look like gigantic stem galls. Normally the galls measure up to 4 cm long by 2 cm in diameter. The color of the fresh galls is a pale green. Heavy infestations of this mite cause noticeable deformations of the petioles and leaves and stunts growth. In places where the growing season is short, the activities of large numbers of this mite have serious effects on the new

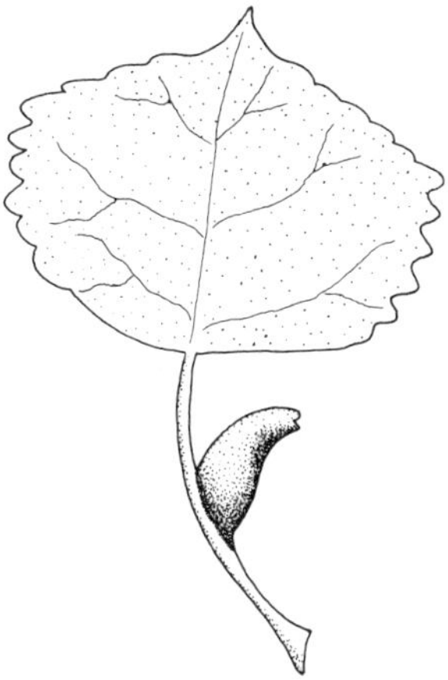

A. Aphid gall of *Pemphigus bursarius*

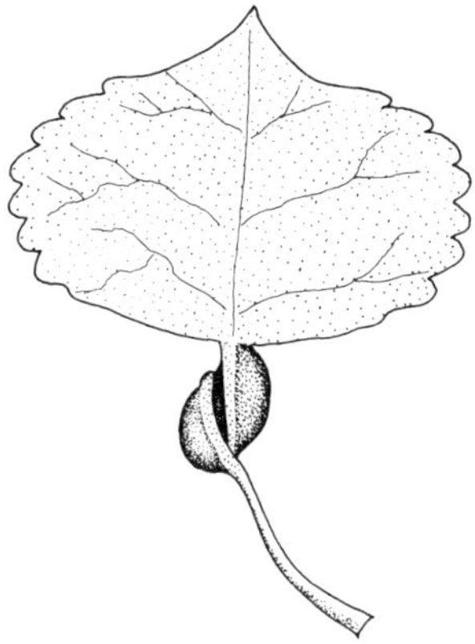

B. Aphid gall of *Pemphigus nortoni*

Fig. 87

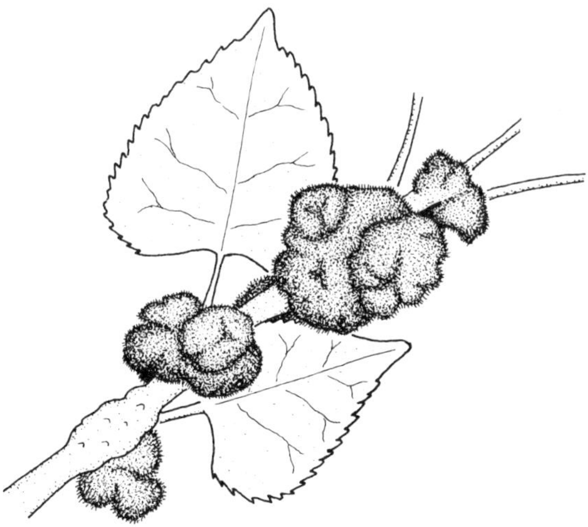

C. Poplar bud gall, *Eriophyes parapopuli*

stems, leaves, and flowers. This mite also occurs on black cottonwood and Frémont cottonwood. It is not uncommon to find heavily infested trees growing side by side with trees not touched by this mite. The poplar bud gall mite is common from Modoc County to Los Angeles County. It is widely known in other western states.

The poplar erineum gall mite, *Phyllocoptes didelphis,* induces yellow-green or pink-flushed, veiny erineum lumps on the dorsal surface of quaking aspen leaves in the spring and early summer. The corresponding hairlined depressions on the undersides of the leaves are reddish when fresh and brown when old. Some galls have hairs that are actually hardened lobes and not the typical hairs found in other erineum galls. These galls measure up to 5 mm in diameter and rise on the dorsal surface of the leaf 2 mm or so. The veins of the leaves show on the convex surface of the erineum galls. Frequently, leaf tissue near the galls blackens as a result of the redirection of energy into the gall. The galls occur singly or in large numbers per leaf. Adults overwinter around the buds. This gall mite is locally abundant, being extremely common in some areas and absent in others.

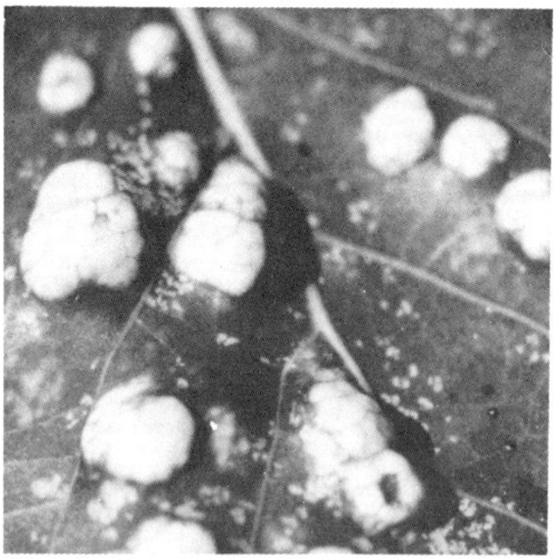

A. The poplar erineum gall as viewed from the dorsal surface of the leaf shows the light yellow bumps induced by the activities of *Phyllocoptes didelphis* on the underside.

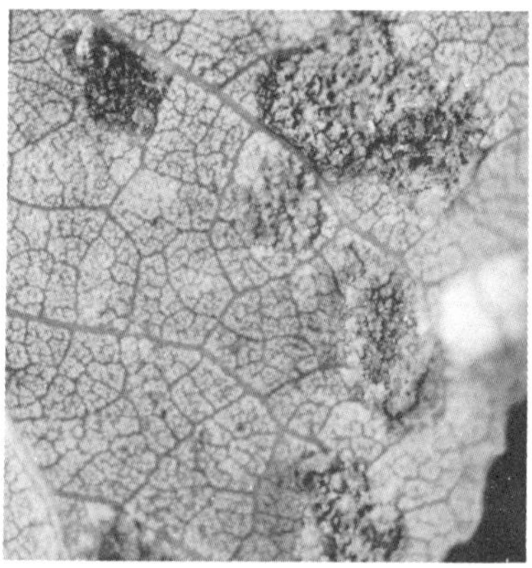

B. From below, the white, pink, or brown hair-lined depressions characterize this mite's galls on aspen.

Fig. 88

Juniper Galls

The junipers of California occur in the higher, cold, mountain regions as well as in the lower arid parts of the state. There are three common junipers discussed herein, even though Munz includes a fourth in the Monterey Bay area. These include the California juniper, *Juniperus californicus,* of the Mohave Desert, southern Sierra, and Santa Lucia Mountains; the Utah juniper, *J. osteosperma,* of the Great Basin country north of Mono County; and the western juniper, *J. occidentalis,* of the high Sierra and the Coast Ranges of northern California.

The galls induced on junipers are largely the results of cecidomyiid midges. The cecidomyiid galls normally occur on the leaves and buds. An unidentified, ping-pong-ball-sized gall occurs on the stems. Like so many other cases where the taxonomy of gall midges has yet to be resolved, two of the four common midges galling junipers remain unidentified.

The juniper cone gall midge, *Oligotrophus betheli,* induces monothalamous, olive-sized, robust, cone-shaped galls in the spring on the leaves of California juniper. It may also occur on Utah juniper as well. The galls form singly or in clusters around the cylindric leaves. The galls measure 1.5 to 2 cm in length. They arise from broad bases and narrow toward the tips, which usually remain closed. Some specimens spread open apically, but rarely all the way. The galls of this midge are characterized by large scalelike bracts, which measure 5-10 mm in width. The bracts do not recurve, but remain erect. When fresh, the galls are green, but turn brown with age. The juniper cone gall has been collected in several localities, particularly from the desert slopes of the Tehachapi Mountains and from the Livermore-Mt. Diablo area of central California. Little is known of the activities of the adult midges.

The juniper urn gall midge, *Walshomyia sabinae,* induces berrylike, urn-shaped, monothalamous, clasping leaf galls in the spring on both California and western junipers. The galls are further characterized by their silver-grey bloom, which develops

A. Juniper cone gall, *Oligotrophus betheli*

B. Juniper urn gall, *Walshomyia sabinae,* in its young stage.

C. In age, the gall splits at the tip, indicating the larvas have pupated and left as adults.

Fig. 90

apparently with age. This latter feature makes the galls closely resemble the normal fruit produced by the junipers. When fresh, the galls are usually greenish-grey, yellow-brown, smooth, glossy, and closed at the apex. With age, the urn galls split at the apex into 3-5 lobes, which recurve and flare out. The resulting hole exposes the larval chamber. Old galls are often rusty brown with hints of the grey-blue bloom, and are noticeably wrinkled. The bases of the galls wrap around the leaves with 3-4 clasping bracts or lobes that measure 2-3 mm wide and long. The entire gall measures up to 7 mm wide to about 10 mm high. The specimens found on junipers in desert areas are fresh in March and April, while the galls on the High Sierra junipers are fresh in May and June. Both young and old galls can be found by the hundreds on a single, small juniper. In July and August in the central Sierra near Truckee closed galls reveal individuals in the last instar larval stage. Open galls rarely contain gall midges, but instead contain the larvas of other insects.

A large burrlike gall occurs on Utah junipers on the eastern, arid side of the Sierra near Loyalton, Chilicoot, and Hallelujah Junction at Highways 395 and 70. These galls superficially resemble those of *Oligotrophus betheli,* except that the bracts are

Fig. 91. Unidentified midge cone gall on juniper looks more burrlike. This gall is often associated with *Oligotrophus betheli.*

narrower, measuring under 4 mm wide by 10 mm in height. The bracts of these galls also differ in that they recurve radically, causing the apex of the galls to be broadly opened. Many of the bracts along the sides of the galls also flare away from the central section of the galls, creating a burrlike appearance. These galls are about 2 cm wide.

A smaller burrlike gall (see fig. 190A) occurs on the junipers in the Tehachapi Mountains. When fresh in the spring, these unidentified galls measure up to 5 mm in diameter. Because no larvas have been found within the galls at this time, it is possible that they may undergo considerably greater growth and may be related to the preceding unknown galls. However, these small galls bear radically recurved bracts that mask the main bodies of the galls. This suggests that the recurving of the bracts is not a process that develops with age.

Maple Galls

Of the four species of maples in California, only one supports a gall organism frequently enough to warrant discussion in this guide. Mountain maple, *Acer glabrum,* is found growing in moist places up 2,700 m in the Sierra Nevada and at about 1500 m in the north Coast Ranges of Del Norte, Trinity, and Siskiyou counties. Its shrublike form combined with the small, three-lobed leaves give the general appearance of a gooseberry instead of a maple. A variety of this low-growing maple is found in the mountains of southern California.

The maple erineum mite, *Eriophyes calaceris,* induces bright magenta, hollow, minute hairs on the upper surfaces of the leaves

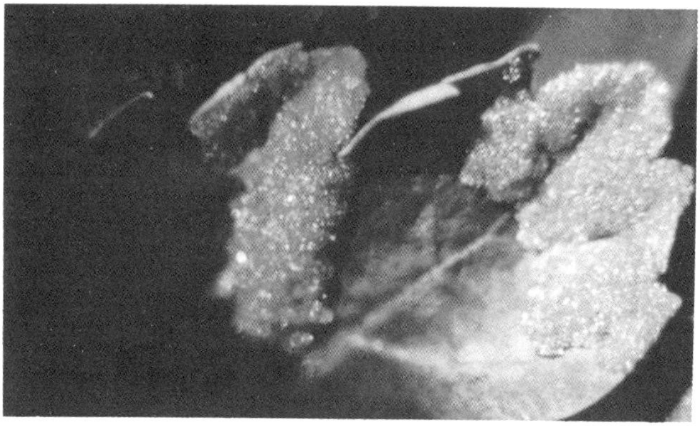

Fig. 92. Maple erineum gall, *Eriophyes calaceris*

in the spring and early summer. While the velvety, red erineum forms broad mats along the toothed margins of most of the affected leaves on a particular tree, other leaves are almost completely covered by the red erineum. The short hairs get their color from the red fluid found inside. The lower surfaces of the leaves are often distorted, but not with the noticeable lumps found with other erineum galls. Numerous individual mites may occupy a single erineum gall. The bright red color makes these trees stand out readily.

A white erineum gall has been reported on the undersides of the big leaf maple, *Acer macrophyllum,* but it appears to be neither common nor clearly identified. However, Keifer has found the same associates of the previous erineum galls here.

Oak Galls

While oaks are of more interest to the gall enthusiast than are any other group of trees or shrubs, they remain a complex, taxonomic problem for the experts. Munz recognizes 16 distinct species with several subspecies and varieties. In California no less than 13 hybrid combinations have been reported. But hybridization is not the only problem in identifying species of oaks. Like other plants, oaks vary in size and leaf shape from one grove to another or from one region to another within the same species. Soil type, slope, and temperature extremes influence the characteristics of the trees and their suitability to being clearly identified and galled. It is clearly beyond the scope of this book to delineate each species of oak in this region. Instead, oaks are divided into the broad subgeneric groups accepted by most botanists, with detailed discussions on some of the more prominent galls associated with this genus of trees.

Subgeneric Groups of Oaks

I. Black Oaks

black oak	*Quercus kelloggii*	San Diego north through Coast Ranges and Sierra to Oregon
coast live oak	*Q. agrifolia*	Coast Ranges, Sonoma County to San Diego
interior live oak	*Q. wislenzii*	inner Coast Ranges, Sierra foothills, Ventura to Shasta and Siskiyou counties

II. White Oaks

valley oak	*Q. lobata*	Sacramento and San Joaquin valleys, inner and middle Coast Ranges
blue oak	*Q. douglasii*	Los Angeles County to head of Sacramento Valley, on slopes of interior valleys
scrub oak	*Q. dumosa*	Nevada to Tehama counties, north Coast Ranges
leather oak	*Q. durata*	middle and inner Coast Ranges from Trinity County to San Luis Obispo County, Sierra from Nevada to Eldorado counties
Oregon oak	*Q. garryana*	Santa Cruz Mts., Marin County to Siskiyou County and British Columbia
deer oak	*Q. sadleriana*	Trinity, Del Norte, and Siskiyou counties north into Oregon
Englemann oak	*Q. engelmannii*	Los Angeles County to San Diego County, inland from the coast

III. Canyon Live Oaks

canyon live oak	*Q. chrysolepis*	many plant communities, widespread Oregon to Lower California, below 2,000 m
huckleberry oak	*Q. vaccinifolia*	Sierra Nevada from Fresno County north to Siskiyou, Del Norte, Trinity, and Mendocino counties, up to 3,000 m
Dunn oak	*Q. dunnii*	San Luis Obispo County, San Bernardino Mts. and edge of the Colorado Desert

With the exception of the Dunn oak, all other species mentioned above have been examined and are included in the descriptions of the galls and the organisms. As mentioned earlier, the gall wasps associated with the oaks are divided among the three subgeneric groups with no crossover of any gall-wasp species from one group to another. Most of the galls discussed under oaks are induced by cynipid wasps. An eriophyid mite incites erineum galls on members of two groups of oaks. There are some midges that gall oaks, but their taxonomy is unclear. Although wasp galls occur on every part of the tree, only those appearing on the flowers, branches, leaves, and petioles are included in this discussion.

Black Oak Galls

Flower Galls. There are three common galls that occur on the catkins or flowers of coast live oak and interior live oak. The flowers of black oak appear to be unaffected.

The gall wasp *Callirhytis congregata* induces large, elongate, fleshy, polythalamous catkin galls. When fresh in April and May, the red and green sausage-shaped galls stand out in their typical clustered groups. Generally the entire flower cluster is engulfed in gall development. At this stage only the anthers protruding from the gall mass give evidence of the muted existence of the flower. The gall, which measures up to 4 cm long by 1 cm in

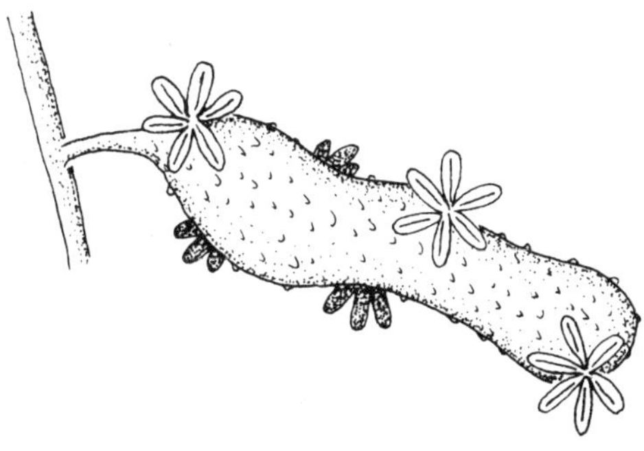

Fig. 95. Cynipid wasp, *Callirhytis congregata*

diameter, assumes the general shape of a potato or sausage. The young growing galls often release a sweet phloem exudate, which attracts ants and other insects. At the peak of this cynipid wasp's cycle, 90% or more of the flowers of an individual tree can be galled. Although adults have been reared in November, little else is known about the insect.

The wasp *Callirhytis serricornis* induces clusters of tiny, kernellike, monothalamous, hard galls of the bisexual generation on the flowers of coast and interior live oaks. The individual galls normally occur in grapelike clusters at the upper end of the catkin. The entire mass measures up to 2 cm in diameter. The individual kernellike galls measure only 2-3 mm long. The galls develop from the pollen sacs at the base of the flowers. The galls are usually brown and obtusely pointed. Exit holes are easily noticed. Males and females emerge in late spring. Females oviposit in the undersides of the veins. In July the galls of the unisexual generation begin development. These green and red galls are round, 1 mm in diameter, and possess a cap or collar on top of each rounded base. Pupation takes place in November after the galls have dropped to the ground. Adults emerge in January in southern California. These parthenogenetic females oviposit in unopened buds of the staminate flowers. Thus, the cycle begins once again.

The two-horned gall wasp, *Dryocosmus dubiosus,* induces glossy, dark brown, club-shaped monothalamous galls on the flowers and new spring leaves of these oaks. These bisexual galls are 2 mm wide by 4-5 mm long. Often two or more galls coalesce creating what appears as a single large gall. The galls generally appear singly or grouped together on a catkin or at the edges of soft new leaves. Galls on the leaves often cause indentations and considerable malformation of the leaves to the point of obliteration of anything leaf like. These galls are further characterized by a vertical groove running the length of each gall. Adults emerge in late April and live for about seven days. Females oviposit in the midrib and lateral veins on the undersides of the leaves. The

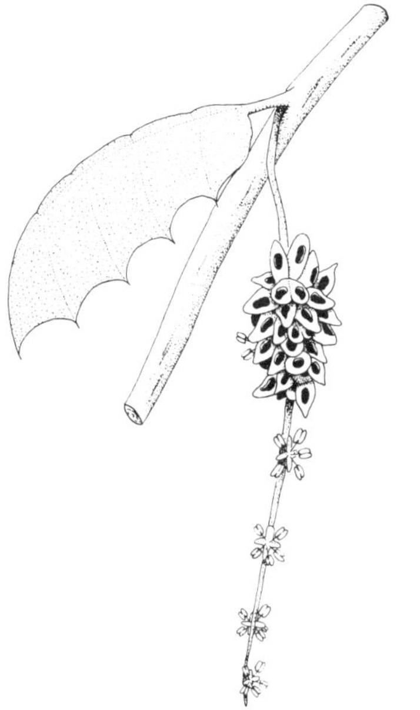

A. Bisexual generation galls of *Callirhytis serricornis*

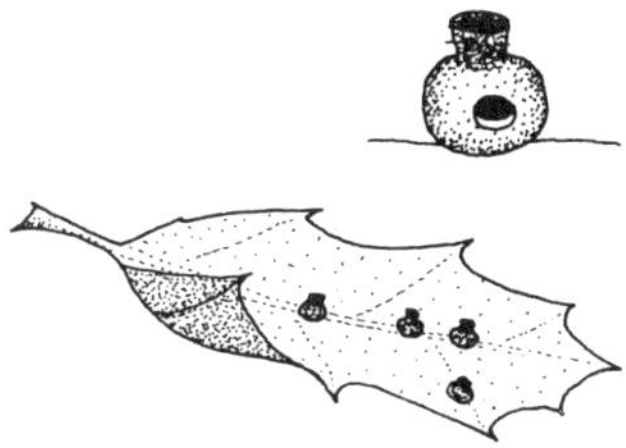

B. Unisexual generation galls of *Callirhytis serricornis*

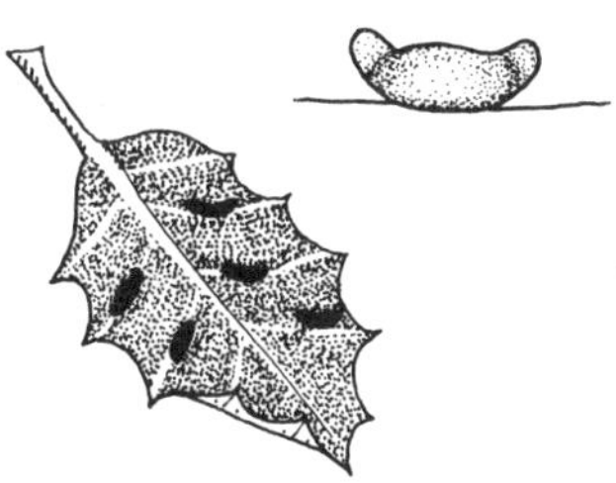

C. Unisexual generation galls of *Dryocosmus dubiosus*

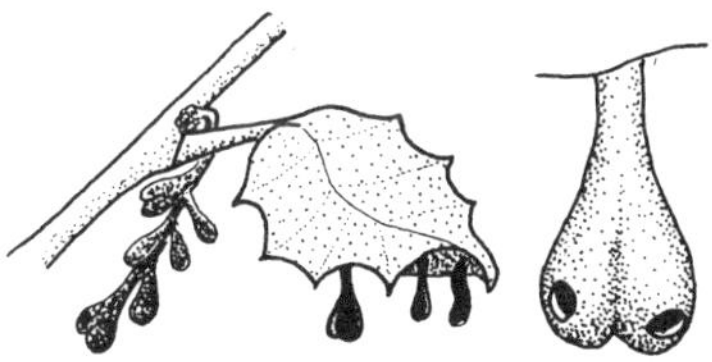

D. Bisexual generation galls of *Dryocosmus dubiosus*

Fig. 97

resulting unisexual galls that develop after the eggs hatch are laterally compressed with a point or horn at each end. Fresh galls are green and measure up to 3 mm long. These galls last on the trees well into the fall and, actually, can be found sparingly throughout the winter and spring. In January and February of 1976 fresh galls were observed by the hundreds. The drought and unseasonal weather of this year may have accounted for this seemingly endless display of galls. On the other hand, different emergence times or a longer larval period than currently acknowledged may have been the reason. The year 1976 appeared to be a peak year in the cycle of this wasp. Observations in the central coastal counties of California revealed that coast and interior live oaks by the thousands were suffering the effects of these galls located on the feeder veins. Leaves were browning out and dropping off the trees by September. At the same time, the population of California oak worm, *Phryganidia californica,* appeared to have peaked in many areas. Although it was suspected that the lack of leaves, resulting from the voracious appetites of the oak worms, could mean disaster for the next generation of these wasps, the oaks quickly recovered, providing the wasps with egg-laying habitats. In spite of the tremendous ecological stress and competition placed on these two insects, they continue to survive with their preferred hosts.

Stem Galls. There are four stem galls that occur on the live oaks and black oaks. Two of these are detachable and two are integral.

The twig club gall wasp, *Callirhytis quercussuttoni,* induces abrupt, integral, globular-shaped polythalamous stem galls on all three oaks in this group. The large, bark-colored galls reach a maximum size of about 5-10 cm in length by 2-4 cm in width. The gall shape can vary from a symmetrical, round form to a lop-sided, knobby, potatolike form. The older galls are often riddled with the exit holes of the gall wasps and intruding parasites and inquilines. These unisexual galls begin their most noticeable development in the summer and continue to grow until the

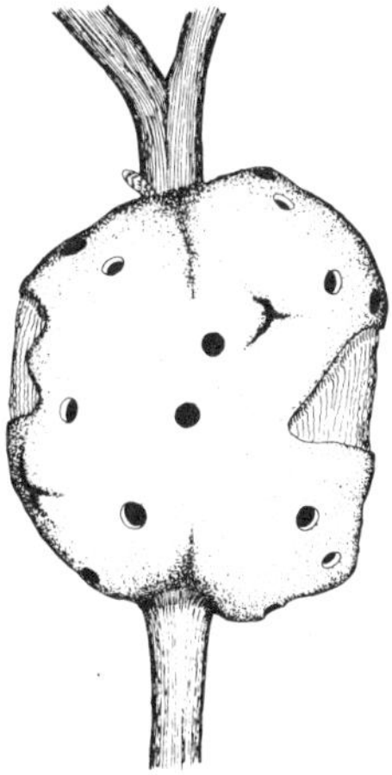

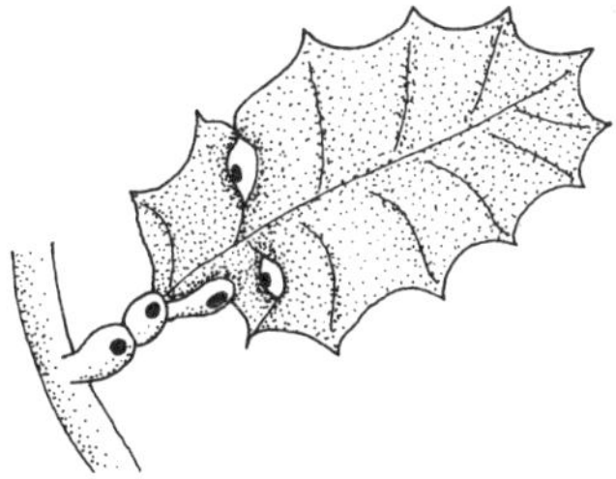

B. Bisexual generation galls of *Callirhytis quercussuttoni*

A. Unisexual generation galls of *Callirhytis quercussuttoni*

Fig. 99

following spring. Larval development terminates in the late spring with pupation taking place in the fall of the second year. Adults emerge from midwinter through March, nearly two years after the beginning of gall formation. Females of this unisexual generation oviposit in the leaf buds. As the leaves unfold in the spring, small (3 mm) green blisters develop on the petioles and leaf veins. From these monothalamous bisexual galls emerge the males and females. The females normally select green twigs 3-5 mm in diameter from the previous year's growth. Later, in the summer, the integral stem galls of the unisexual generation begin development.

The activity of development of the unisexual stem galls usually causes the death of the branches beyond the galls. The extensive disruption of the flow of essential nutrients to the outer twigs and leaves, as caused by the galls, is usually the fatal blow. In an oak woodland, where many natural controls exist, this wasp rarely causes any serious harm to an established tree. However, isolated ornamental live oaks can suffer serious damage after succeeding years of attack without the normal complement of parasites and predators to keep the gall wasps in check.

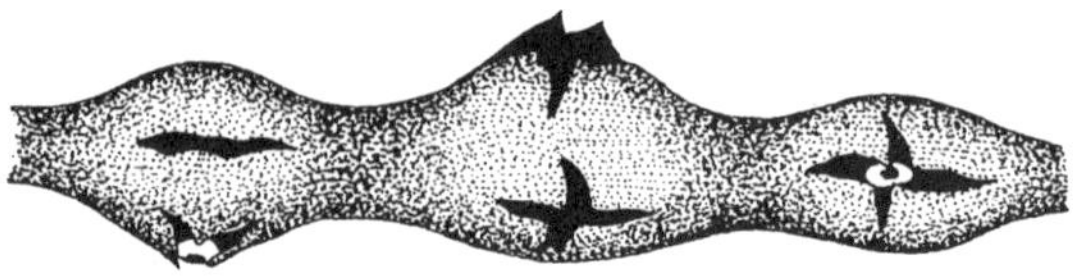

Fig. 100. Cynipid wasp gall *Callirhytis perdens*

The ruptured twig gall wasp, *Callirhytis perdens,* induces polythalamous, gradually swollen stem galls that rupture, expelling the larval chambers. The galls measure up to 7 cm in length and 2.5 cm in diameter. While the galls are bark-grey in color, they are noticeably characterized by the numerous slender black fissures. These cracks in the gall's surface develop as the larvas inside the chambers mature. A dozen or more glossy, laterally flattened, tan larval chambers are slowly forced out of the gall through the fissures. At this time the galls are often still green and soft, but by autumn they become brown and hard. The larval chambers fall to the ground where they seemingly vanish amongst the forest humus. Once on the ground, the larval chambers can easily be mistaken for seeds by birds and mice. They are also carried off by ants as are the galls of *Dryocosmus dubiosus.*

An interesting advantage of this rupturing mechanism is that the adult need only chew its way out of the gall chamber and not the hard tissue of the gall proper. On the other hand, there is increased susceptibility to predation once the larval chambers have been expelled from the galls. Like other integral stem galls, those of *C. perdens* cause damage to the twigs beyond the galls. Little is known about the adult insect.

The live oak gallfly, *Callirhytis pomiformis,* induces two distinct galls belonging to two separate generations of this species. The unisexual summer stem gall is detachable, round, to 3 cm in diameter, polythalamous, and covered by numerous hard, pointed projections. These prickly protruberances measure

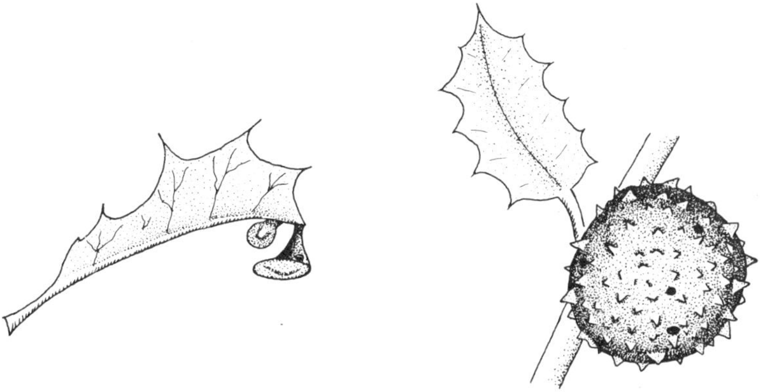

A. Bisexual generation galls of *Callirhytis pomiformis*

B. Unisexual generation galls of *Callirhytis pomiformis*

Fig. 101

about 2 mm in length. When fresh, the galls are succulent with green and red tones. With age, the galls fade to light brown. The mature galls are lightweight, but solid and composed of pulpy, balsalike material. The larval chambers are located centrally. The parthenogenetic females emerge from the galls between February and March. They oviposit in the unopened leaf buds. As the leaves unfold, the alternate generation of bisexual galls begins development on the undersides of the leaves. These galls appear singly or in groups of two or three and may remain on the stems of the trees for two or more years. The mushroom-shaped galls measure about 5-7 mm wide and high. When fresh, these galls resemble goblets or mushrooms with light yellow, green, red, and pink tones. As the galls reach full growth, the cuplike portion flattens out. With age, the galls turn beige. These monothalamous galls develop within two weeks after oviposition. The stalks are solid, while the heads of the galls are comprised mainly of the larval chambers. Males and females emerge in mid-May to early June. Robert Lyons, the researcher who discovered this unique relationship between alternating generations of this wasp, found evidence that suggested some of the parthenogenetic females laid eggs which produced only sexual females. Other parthenogenetic females laid eggs which produced only males in

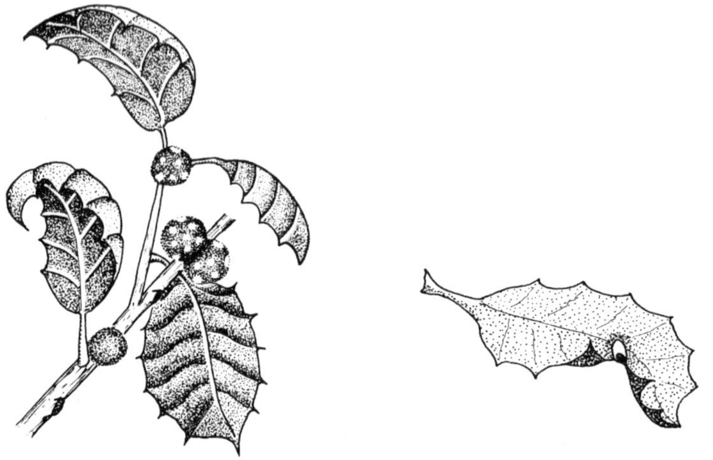

A. Unisexual generation galls of *Callirhytis agrifoliae*

B. Bisexual generation galls of *Callirhytis agrifoliae*

Fig. 102

their respective galls. Lyons estimated that more than 75% of the bisexual galls were parasitized.

The wasp *Callirhytis agrifoliae* induces green, purple, beige, and white mottled, round, monothalamous galls attached to the stems at the terminal leaf axils. The minutely pubescent galls measure from 6-9 mm in diameter. Development begins in early summer and continues for several months. Around October and November the now brown galls drop to the ground by the thousands. Many of these galls are taken by squirrels, birds, and people interested in making necklaces. In February parthenogenetic females emerge to oviposit in leaf buds. As the leaves unfold, tiny 2 mm-long, one-celled blister galls develop on the leaves, in aborted buds, and on the petioles. They develop rapidly, with the adults of this bisexual generation emerging in April, 7-8 weeks after egg deposition by the parthenogenetic females. While the blister galls are difficult to locate under superficial inspection, the round, fuzzy bud galls are extremely common and easy to find on the trees or the ground.

Leaf Galls. There are four common leaf galls that occur on members of the black oak group. These oaks clearly do not support the diversity of galls that are found on the white and canyon live oaks.

The wasp *Dryocosmus minusculus* induces tiny 1 mm in diameter, beadlike galls on the upper surfaces of coast and interior live oaks and black oaks. When fresh in the summer, these monothalamous galls are light yellow with dark red or purple centers. As the galls age and lose their color and moisture, they assume the appearance of lilliputian pumpkins with several furrows radiating toward the depressed centers. In the fall most of the galls drop to the ground where they blend among the leaves. Adults emerge in April. In some areas, one finds fresh galls at just about any time of the year.

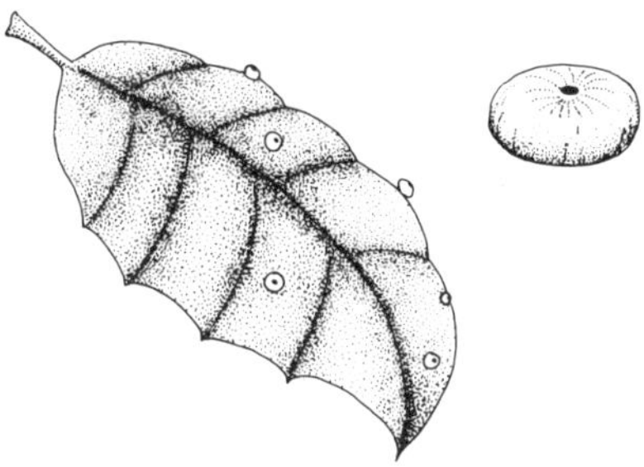

Fig. 103. Cynipid wasp gall *Dryocosmus minusculus*

The live oak petiole gall wasp, *Callirhytis flora,* induces large, hard, smooth-surfaced, irregularly shaped galls that often encompass the entire petiole and/or part or all of the midrib vein of the leaves of these oak trees. These polythalamous galls develop in the spring on the new leaves and petioles and measure up to 5 cm long by 1 cm in diameter. These integral, hard galls show more on the ventral side of the petioles and leaves than on the upper sides, which the galls often distort due to their growth pattern. When fresh, the galls are green, but turn light, glossy brown with age. From these galls emerge males and females in May and June. Impregnated females oviposit in young acorns. In

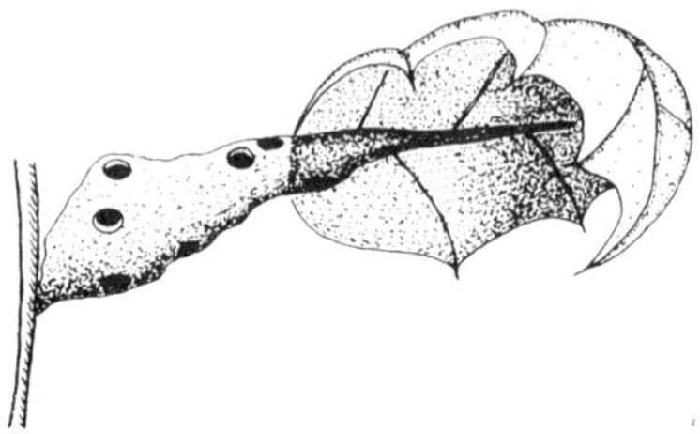

A. Bisexual generation gall of *Callirhytis flora*

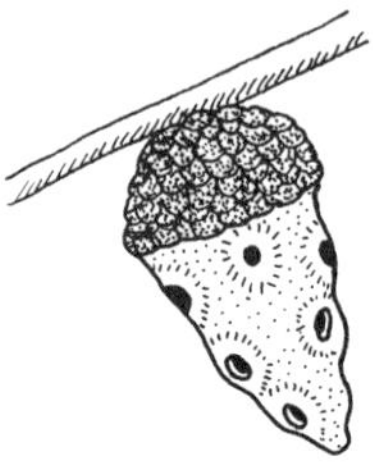

B. Unisexual generation gall of *Callirhytis flora*

the fall the aborted galled acorns drop to the ground, where the unisexual generation larvas remain in a prepupal stage for about 1.5 years. The parthenogenetic females emerge from the acorn galls (that are not eaten by squirrels, jays, and deer) in February and March to oviposit in the petioles and midrib veins. This species is extremely common in the central Coast Ranges and the Sierra foothills.

The wasp *Callirhytis perfoveata* induces round, monothalamous, integral leaf galls in the spring on all species in this group. The galls are situated almost equally on both sides of the leaves. When fresh, the galls are succulent and green; with age they turn brown and become very hard. These smooth, glossy galls are usually less than 1 cm in diameter. In cross section one sees empty cavities next to the centrally located larval chamber in these galls. Adults emerge in late April and May. Thus far, any alternate generation that may exist remains undetected.

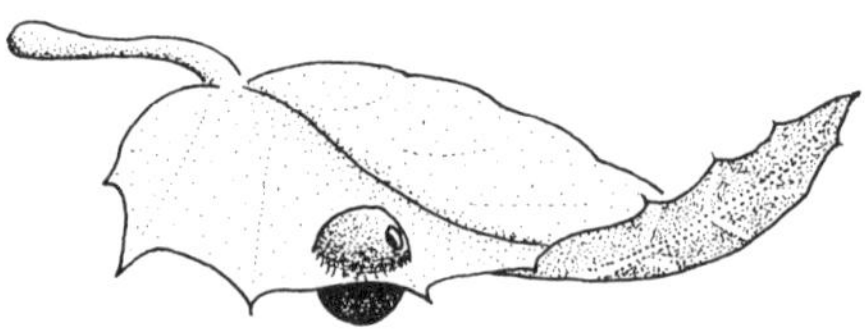

C. Cynipid wasp gall *Callirhytis perfoveata*

Fig. 104

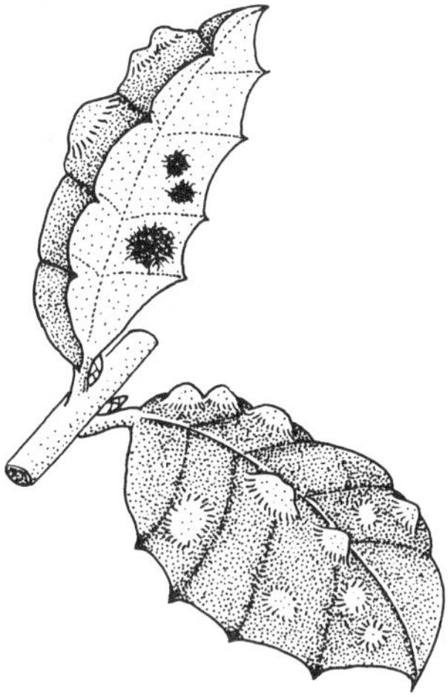

Fig. 105. Erineum gall, *Eriophyes mackiei,* on coast live oak

The live oak erineum mite, *Eriophyes mackiei,* induces rusty brown, fuzzy mats on the undersides of the leaves of black and canyon live oaks. This crossover from one oak subgenus to another is unique among the oak-galling organisms. The rusty-haired pockets have corresponding convex lumps on the dorsal side of the leaves. Some leaves are so thoroughly infested with this mite that the entire undersurfaces of the leaves are covered with the easily noticed erineum, and the upper surfaces are considerably distorted. On the other hand, one can find erineum pockets that are as small as 1-2 mm in diameter. The mites generally overwinter in the erineum pockets, while a few may seek the shelter of buds. New leaves are infested in the spring and early summer. The short, rusty hairs are merely extensions of normal plant pubescence. The galls of this mite are found throughout the range of the host oaks.

White Oak Galls

The white oaks are notable for hosting the most colorful, ornate, and largest cynipid galls in our region. Although a few species gall flowers and acorns, the majority of galls occur on the leaves and stems. Over 50 species have been recorded on leaves and stems, while many others remain unidentified as new species or as alternate generations to species already known.

Stem Galls. While the diversity of form and color found in the stem galls of white oaks can serve only as a prelude to the magnificence and wealth of leaf galls, they are, nevertheless, unique and fascinating.

The irregular spindle gall wasp, *Andricus chrysolepidicola,* induces integral, polythalamous, spindle-shaped stem galls on

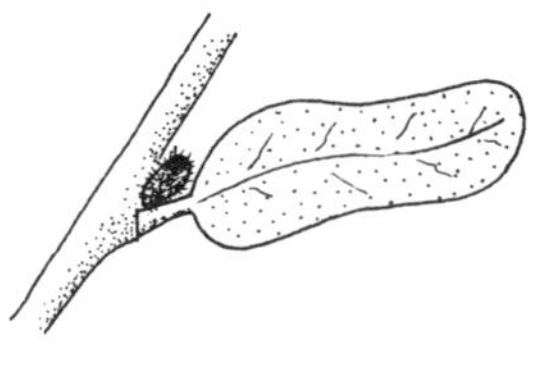

A. Bisexual generation gall of *Andricus chrysolepidicola*

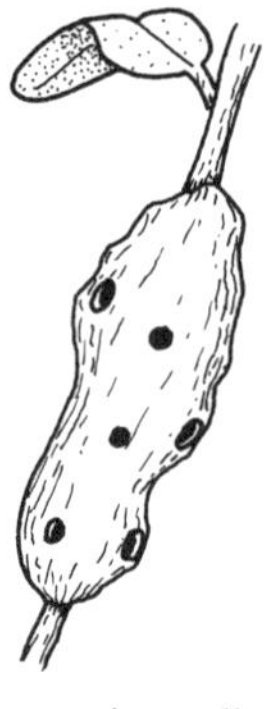

B. Unisexual generation gall of *Andricus chrysolepidicola*

Fig. 106

blue, valley, scrub, and leather oaks. The galls range in size from 3-7 cm long by 1-2.5 cm in diameter. When fresh in the spring and early summer, the galls are soft and easy to cut open. By late summer the smooth, sometimes knobby galls are extremely hard. Apparently, the galls of this species cause little damage to outer plant parts, unlike other integral stem galls that seriously interrupt the flow of nutrients. These abruptly swollen galls are the same color as normal bark. Adults chew their way out of the galls between April and May of the following year. The numerous tiny exit holes scattered over the surface of the galls are evidence of the mass emergence of adults. The bisexual generation of this species develops in 2 mm-long, ovoid bud galls, which are covered by a fine pubescence.

The California gallfly, *Andricus californicus,* induces the largest cynipid galls in California. When fresh in the spring, these **oak apples** are succulent, green or red, smooth and glossy. These

Fig. 107. A large cluster of galls by *A. californicus*

so-called "apples" taste very bitter due to the high concentration of tannic acid. By late summer the galls turn creamy-white and normally bear blunt, short knobs or wartlike protruberances. By fall, the galls are light brown. With advancing age, the galls may become covered with a black, sooty mold. The galls are round to potato-shaped and measure up to about 12 cm long by 7 cm in diameter. They occur singly or in clustered groups of four or five. Under the conditions of clustered growth, the galls may appear more like pears in shape. Although easily detached from their stem connections, the galls of this wasp may remain on the tree as long as four years after formation. This species is known only from parthenogenetic females that emerge in October and November; no males are known for this species. Pupation and emergence of the females may not take place the same year as the development of the galls, but a year or more later.

These large-bodied wasps (3-4 mm) oviposit in late fall. Gall development does not commence until spring. Between March and May, depending on the geographic region, the bark of oak branches ruptures under the pressure of the developing gall inside. Frequently a piece of bark forms a lid, with the young gall clearly visible from an angle. When these galls first make their appearances, they measure about 1.5 cm in diameter. They usually reach full development within about two months. A second brood of galls begins to appear in July, but these are from eggs laid at the same time as those of the spring crop of galls. The reason for this delay and the two different broods is not clear. Females from both broods appear to emerge at about the same time.

Old galls yield a variety of chalcid and braconid wasps, as well as filbertworms, and several other inquilines, parasites, and hyperparasites. Woodpeckers frequently dig into these galls in search of these morsels. It has been reported that these oak apples were a source of material for an eyewash used by southern California Indians. The oak apples of the California gallfly are extremely common, ranging from Washington southward into Mexico. Because these galls often occur by the thousands on a single tree, they may represent a serious nutritive drain on the host plant. This species occurs on all white oaks mentioned earlier.

The wasp *Disholcaspis canescens* induces monothalamous, detachable, round, silvery, slightly mottled stem galls on blue and scrub oaks. The silvery-grey bloom that covers the surface easily rubs off. The globular galls measure about 2 cm in diameter. Scattered tubercles are found on the surface of most galls. Females emerge in February and March. Abandoned galls remain on the host trees for some time after the adult insects have left. Little else is known about the life history of this species.

The coral gall wasp, *Disholcaspis corallina,* induces round, monothalamous, orange, yellow, and red detachable stem galls in the summer on blue oak. The prime characteristic of these galls, which measure 1 cm in diameter, is the numerous blunt, clublike projections 2 mm long, scattered over the minutely pubescent surface. It is this character that creates the resemblance to a south seas coral head. With age, the galls lose their colors, turn black with a sooty mold, and become extremely hard. They occur singly or in tight clusters of 3-4 at the bases of leaf petioles near the tips of branches. Adults emerge in the fall. Even though this species induces one of the more attractive cynipid galls, little is known about its life.

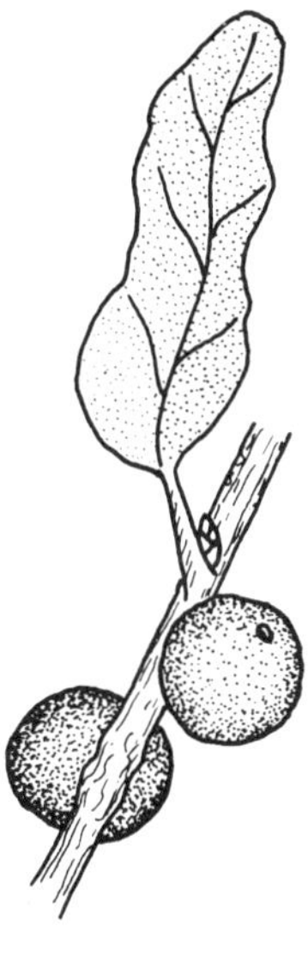

A. Cynipid wasp gall *Disholcaspis canescens*

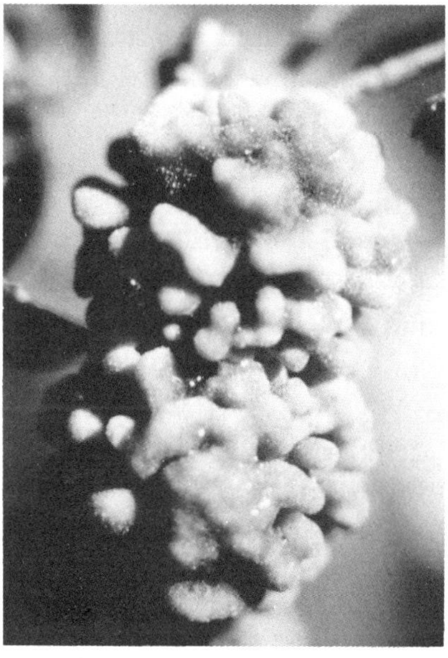

B. Coral gall *Disholcaspis corallina*

Fig. 109

The wasp *Disholcaspis washingtonensis* induces round, detachable, monothalamous stem galls on almost all white oaks. The 8 mm in diameter galls have a mealy granular texture. They occur singly or in close groups, with their basal attachment points often close together. Some forms of these galls have short necklike stalks at the bases. The larval chamber is centrally located in the otherwise solid gall. With age, the brown galls develop a shallow, wrinkled texture to the surfaces. Adults emerge in November. Old galls remain on the trees for a year or so after emergence of the adults. Beetles feed on the old gall tissue, hollowing out the interior to the point where the skin is only a skeletal reminder of the once hard, solid gall.

The wasp *Besbicus conspicuus* induces a detachable, monothalamous, solid stem gall on valley oaks. The outer shell of the 8-10 mm in diameter galls is wrinkled and about 1 mm thick. The inner cell tissue is a rich chocolate brown. When fresh in the summer, the galls are dominated by red tones. The galls occur singly or in clusters that squeeze together, forcing the sides to flatten and taper toward the bases. Single specimens are often perfectly round. I have not found any with a distinct neck at the base. Galls are fully grown by August. With age, they fade to grey-brown, bark-colored tones. Adults emerge in late winter and early spring. This gall is similar to *D. washingtonensis.*

The honey dew gall wasp, *Disholcaspis eldoradensis,* induces round to elliptical, monothalamous, detachable stem galls on valley, scrub, leather, and Oregon oaks in the summer. The galls are generally elliptical with a flat or slightly convex top. The sides are glossy, wrinkled, and light brown. The tops of the galls are dull, nonglossy, fissured, rough, and dark brown. The galls usually have a nippled attachment that fits into the stems; when pulled off the stems, the galls leave prominent holes where the pointed nipples were attached. The galls reach their maximum size of about 8 mm in diameter by September. While the larvas are actively feeding, the galls release a sweet phloem exudate, which attracts ants and bees. The galls remain on the trees for

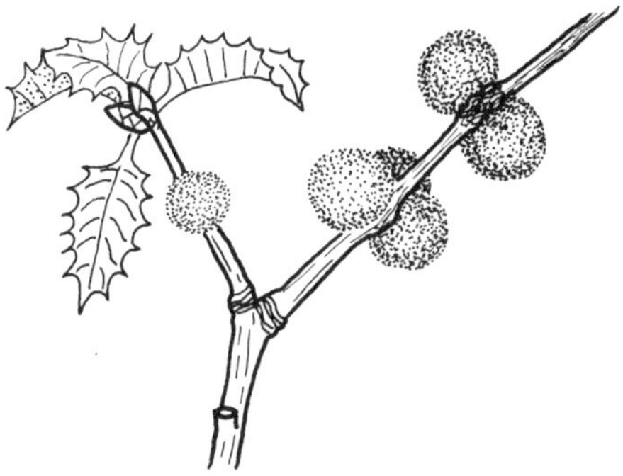

A. Cynipid wasp gall *Besbicus conspicuus*

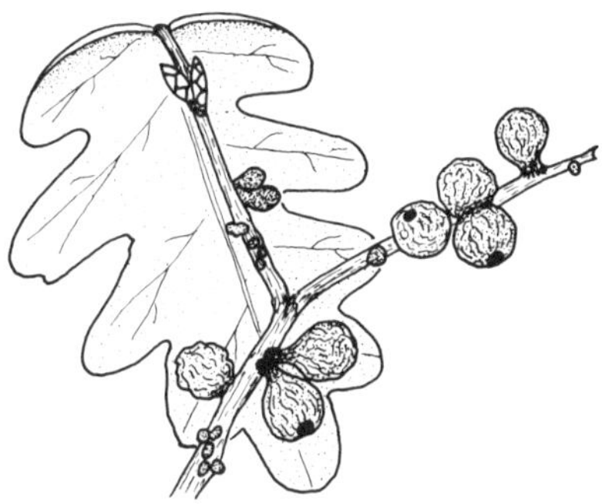

B. Cynipid wasp gall *Disholcaspis washingtonensis*

Fig. 111

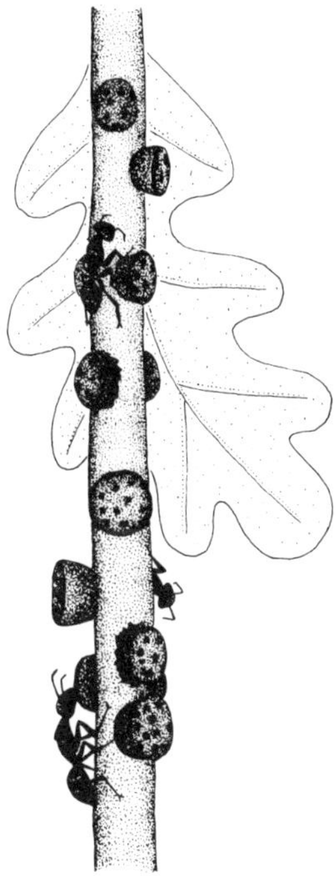

C. Cynipid wasp gall *Disholcaspis eldoradensis*

several years as evidenced by the presence of lichens on many specimens. Gall development occurs in rows, tight clusters, or singly. When several individuals grow in a clustered fashion, the galls can inflict damage on the outer branches and leaves. Adults emerge through holes in the sides of the galls in January and February.

The beaked twig gall wasp, *Disholcaspis plumbella* (see front cover), induces strikingly beautiful monothalamous, round, lumpy-surfaced stem galls on blue, scrub, and leather oaks in the late spring and early summer. The characteristic beak protrudes at the opposite end of the point of attachment and measures up to 6 mm in length. This beak is normally recurved and sharply pointed. The main body of the gall measures up to 15 mm in diameter. This solid gall has a central larval chamber. The galls appear in two color forms: one a solid bright yellow, the second a deep wine red with bright lemon-yellow, round lumps. The latter spotted yellow color phase is particularly noticeable among the branches of the host trees. Gall development begins in May in some areas and reaches full growth by August. The galls remain on scrub oaks for several years. With age, the galls lose their bright colors and become grey to black. Adults emerge in November and December through holes in the sides of the galls. If the hole is situated in just the right place, one can liken the gall to the rounded head of an eagle with its dark eye and curved beak. This wasp and its gall are extremely common in Sonoma County and the area of the Pinnacles National Monument.

The clasping twig gall wasp, *Disholcaspis prehensa,* induces monothalamous, mushroom-shaped stem galls on scrub and leather oaks. The sides of these galls are smooth, glossy, and flaring, the tops rough, fissured, and nonglossy. The broadly clasping base of the gall, which is often wrapped far enough over the stem to completely conceal it, is the chief character of this gall. The sides rise to a constricted neck, upon which sets the pitted cap. These galls measure about 10 mm wide at the base with caps to about 5 mm wide and a total height of 7-10 mm. The

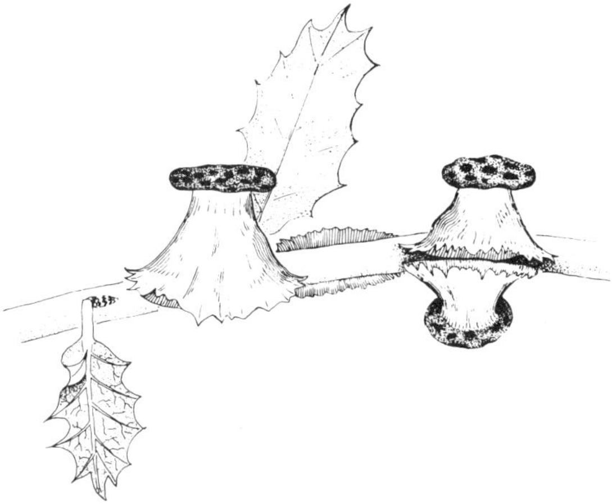

Fig. 113. Clasping twig gall, *Disholcaspis prehensa,* releases a sweet phloem exudate which attracts ants and bees.

sides of the galls facing the sun are usually red, while the shaded sides are beige, yellow, or light green. These galls occur singly or in rows on opposing sides of the stems. Another form of this gall is found in the Mt. Diablo area, and appears more phallic in shape with conical, pitted caps, little or no neck constriction, and the usual flaring, clasping base. These galls are often taller than the mushroom-shaped forms, reaching a height of 12-14 mm. Both forms of this wasp's gall begin development in spring and reach maturity by early summer. Fresh galls exude copious quantities of sweet phloem sap, which attracts hordes of ants. When removed from the stems, the galls leave deep pits where the basal nipples were attached, holding the galls in place. The elliptical larval chamber is located at the base of the otherwise solid gall. Adults emerge through holes in the sides in late winter and early spring. Smaller inquilinous insects occasionally pupate in and emerge from the caps at another time.

Leaf Galls. The leaf galls found on white oaks reflect the ultimate in design, color, and sophistication in adaptation within the insect world. There are over 30 species of wasp galls described as occuring on the leaves of white oaks along the Pacific slope.

Many others remain unidentified. Those discussed herein represent a small segment of the variety of leaf galls to be found. Although abundance varies from one geographical region to another, I have described those most likely to be found generally.

The crystalline gall wasp, *Andricus crystallinus,* induces caterpillarlike masses of bristly-haired galls on the undersides of the leaves of blue, scrub, leather, and Oregon oaks in the late spring and early summer. Actually, the caterpillarlike mass is composed of numerous individual tongue-shaped, spherical, monothalamous galls arising from separate points of attachment. The individual galls measure 12-14 mm high by about 7 mm in diameter. The combined mass may reach 3.5 cm long by 2.5 cm wide and high. Each individual gall normally has a recurved beak and a sparse coating of crystalline hairs. While the gall body may be bright, solid red, the brittle hairs may be white, pink, red, brown, or some shade thereof. The beaks of the galls often protrude through the solid-appearing mass of crystalline hairs. Occasionally, one can find separate galls without the brittle hairs. These bare galls appear to be aborted for some reason, as I have rarely found them to contain fully developed larvas. The early death of the larva probably prevents the gall from developing normally. The larval chamber occupies the entire length of the small gall. In late winter, adult females emerge from these galls, chewing exit holes near the beaks. They reproduce and oviposit in leaf buds. About 17-20 days after oviposition, noticeable green, conical, monothalamous galls with cottony fibers develop on the upper surface of the leaves. The fibers are generally longer than 2.5 mm and tend to project laterally. The gall measures about 2.5 mm long by 1 mm in diameter. These galls, which bear the bisexual generation, are curved. Males and females emerge by March to mate; females oviposit in the tissues of the undersides of the leaves. The adults of both generations are morphologically different. Little damage seems to be done to the leaf tissues of the host trees, even though the galls of the unisexual generation are massive and often cover the entire leaf surface.

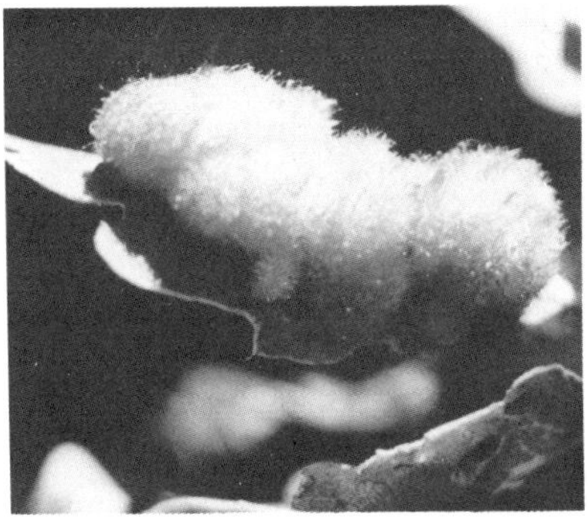

A. Mature unisexual generation galls of *Andricus chrystallinus*

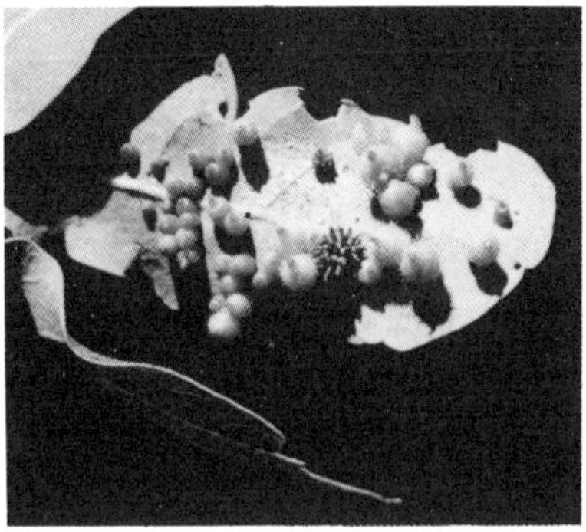

B. Immature or aborted unisexual galls of *Andricus chrystallinus*

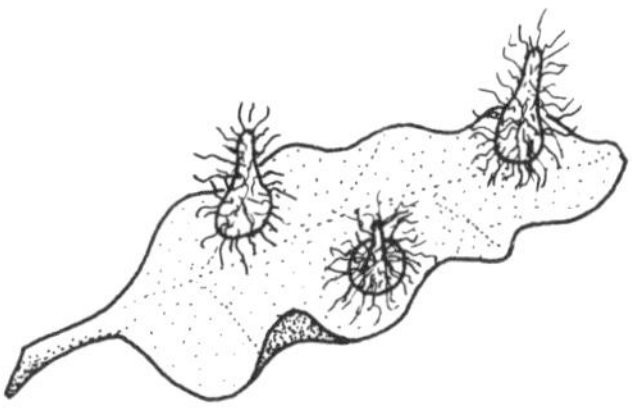

C. Bisexual generation galls of *Andricus chrystallinus*

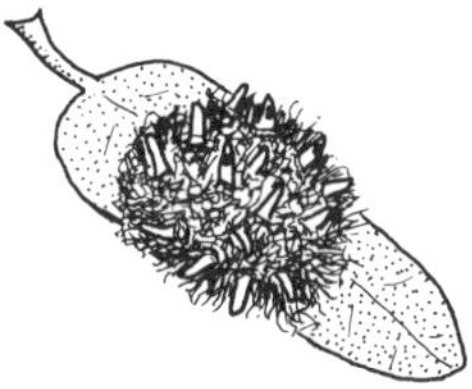

D. Unisexual generation galls of *Andricus chrystallinus*

The cynipid wasp *Andricus brunneus* induces monothalamous, round, pubescent galls on the midrib veins on the undersides of the leaves of most white oaks. The galls occur singly or in clusters. They are dull lavender to tan and are covered with short, cream-colored, matted hairs. The galls measure up to about 6 mm in diameter. The outer wall of the

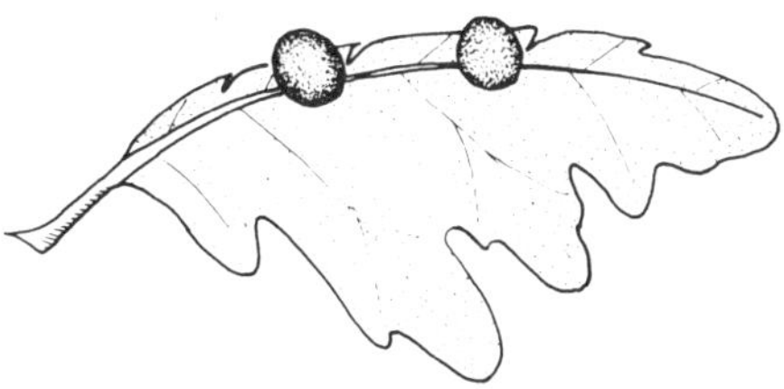

E. Cynipid wasp gall *Andricus brunneus*

Fig. 115

solid gall is thick and hard enough to bend the average razor blade. The larval chamber is located centrally. It is surrounded by lavender-colored cell tissue. The unisexual adults emerge in the fall. No alternate generation is known. The easily detached galls cause little damage to the leaves.

The red cone gall wasp, *Andricus kingi,* induces monothalamous, detachable, cone-shaped, bright red galls on both sides of the leaves of valley and Oregon oaks in the summer months. The cone rises from a flared, cup-shaped base that is narrowly attached to the leaf. The apex of the gall is blunt. The gall measures up to 5 mm high by 3-5 mm wide at the base. The galls of this wasp are never red and white striped as are those of *A. atrimentus.* Dozens of *A. kingi* galls can be found situated on the veins of a single leaf. They generally occur singly, but frequently close together. Under magnification a fine pubescence becomes apparent over the entire gall surface. The larval chamber is large, occupying most of the gall by early fall, when the gall drops to the ground. Parthenogenetic females emerge through holes in the tips of the galls in February. The identification of the bisexual generation of this wasp is controversial. One researcher has associated the females from the red cone galls with the spring galls and the males and females that emerge from these galls, which are usually referred to as *Liodora dumosae,* as the alternate bisexual generation. But, the correct taxonomic use remains unclear for the time being.

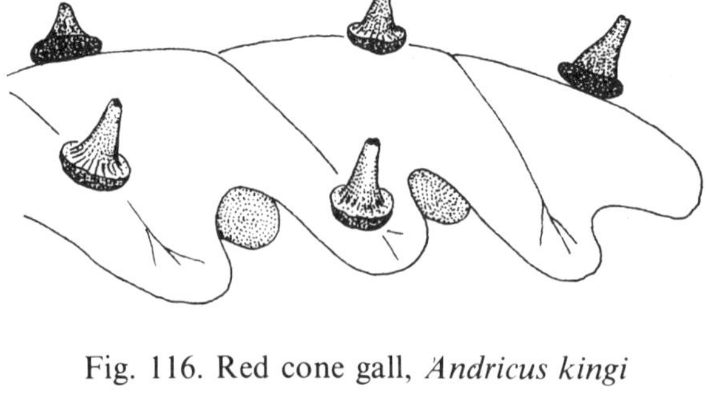

Fig. 116. Red cone gall, *Andricus kingi*

The saucer gall wasp, *Andricus gigas* (see back cover), induces small, 3-4 mm wide, flat, disc-shaped, monothalamous galls on the upper surfaces of the leaves of blue, scrub, and Engelmann oaks in the summer. The saucerlike galls are narrowly attached at

A. Bisexual generation galls of *Andricus gigas*

B. Unisexual generation galls of *Andricus gigas*

Fig. 117

the base and have thin, flaring edges that are either smooth or toothed (**crenate**). The larval chamber is designated by a small lump in the center of the otherwise depressed upper surface of the gall. The surface of the gall is nonglossy. The color of these unisexual generation galls varies from shades of pink to red, purple, and brown. A short, matted pubescence is seen under magnification. The galls occur singly or in large groups of as many as 25-30 on a single leaf. Adult females emerge between December and February. By April the bisexual generation galls are well developed on staminate flowers and leaves. These galls are conical, tan, and about 2-3 mm in height. Males and females emerge from these monothalamous galls to start the cycle once again. The galls of the unisexual generation can be confused, after cursory examination, with the galls of *Andricus parmula*.

The wasp *Andricus fullawayi* induces kernel-sized galls that are covered with a dense mat of long, straw-yellow hairs. These monothalamous, detachable, fluffy-haired galls measure 5-8 mm in diameter. The galls occur singly on the midrib veins on the

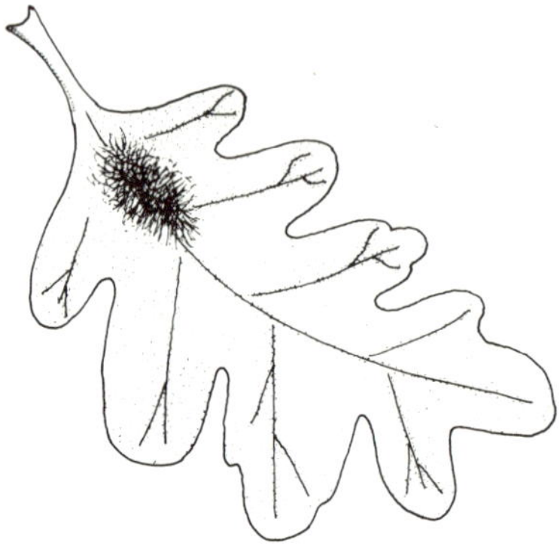

Fig. 118. Cynipid wasp gall *Andricus fullawayi*

undersides of the leaves of most white oaks. The long, soft, straw-yellow hairs distinguish this gall from all other hairy leaf galls on white oaks. These unisexual summer galls are vacated by females in March and April. Little else is known about this gall organism and its gall.

The disc gall wasp, *Andricus parmula,* induces monothalamous, disc-shaped, glossy, detachable galls on both sides of the leaves of most white oaks in the summer. The gall, which measures 2-3 mm in diameter, has a narrow basal attachment, a smooth edge, and a glossy dorsal surface. The surface is streaked with red-brown tones against a faint yellow. The upper surface of the gall is marked by a slight knob or **umbo** over the centrally located larval chamber. These galls have more of a flat appearance to them than the concave galls of *A. gigas.* The galls of *A. parmula* occur singly or in small clusters, but are not confined to the main veins. Adults emerge in April.

The sun burst gall wasp, *Andricus stellaris* (see back cover), induces convex, round, monothalamous galls with radiating crystalline projections on the undersides of the leaves of blue and Oregon oaks. These summer galls measure 3-4 mm in diameter, including the crystalline fibers, which are brittle and about 1 mm long. These radiating fibers spread outwardly in overlapping layers. The center of the gall is covered with short protuberances as seen under magnification. The galls are white, pink, and rose-red. These galls are further characterized by a hollow cavity located above the larval chamber of each gall. The galls occur singly or in clustered groups. Females emerge in the spring. An alternate generation has not yet been determined. When looking down on the galls of this species, it is not difficult to visualize stars or the sun with radiating light rays. This species is common in Sonoma and Mendocino counties and in the southern Sierra Nevada.

The cynipid wasp *Andricus pattersonae* induces disclike thin, flat, detachable, monothalamous galls on the undersides of the leaves of most white oaks. These summer galls measure 7-9 mm in diameter. They have smooth or crenate margins and are narrowly attached just under the central larval chamber. The upper surface of the galls is usually marked with a central depression. No part of these galls appears more than 1.5 mm thick; the edges are thin and fragile. Fresh galls range in color from bright lemon yellow, to mint green, to beige. The larval chamber is ovular in vertical cross section. Several galls often occur on the same leaf. However, it is not uncommon to find only

A. Disc gall *Andricus parmula*

B. Cynipid wasp gall *Andricus pattersonae*

Fig. 119

a single gall. Although they cause some discoloration on the dorsal surface of the leaves, these galls apparently are not responsible for any serious damage to the host leaves. Gall development begins in midsummer with adult emergence taking place the following spring.

The wasp *Andricus confertus* induces convoluted, midrib clusters of finely pubescent, detachable, monothalamous galls on the undersides of valley oak leaves. The brainlike gall mass is composed of individual, triangular-shaped galls measuring 3-5 mm in height. They develop in close proximity to each other, creating the general appearance of the convolutions. The entire mass measures up to 14 mm in diameter when fresh and succulent. Fresh galls are red to red-brown. Normally there is but one gall mass per leaf. Development begins in early summer. Adult females generally emerge the following spring, but in some cases the females do not emerge until a year or more after gall development.

The woollybear gall wasp, *Sphaeroteras trimaculosum,* induces round to ovoid, monothalamous, solid, bristly galls on the midrib and lateral veins on the undersides of valley, scrub, blue, and Oregon oak leaves. These detachable, rusty-brown galls are characterized by the stiff, crystalline, erect hairs, which cover the entire surface of the galls uniformly. The galls are 3-4 mm in diameter and occur singly or grouped on the leaves. When several of these galls develop close together, they resemble a woollybear caterpillar. Only parthenogenetic females are known to emerge from these summer galls. Emergence has been recorded at various times of the year, but based on my observations, I believe that most of the adults emerge in late winter and early spring. No alternate generation is known.

The jumping oak gall wasp, *Neuroterus saltatorius,* induces tiny (1 mm in diameter), round, monothalamous, detachable brown galls on the lower surfaces of blue, valley, scrub, and Oregon oak leaves. These late spring galls occur singly and often in large numbers per leaf. The glossy, smooth brown galls make

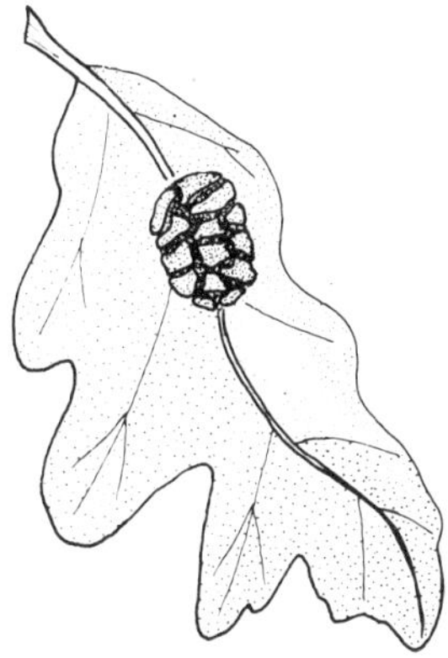

A. Cynipid wasp gall *Andricus confertus*

B. Woollybear gall, *Sphaeroteras trimaculosum*

C. Unisexual generation galls of *Neuroterus saltatorius*

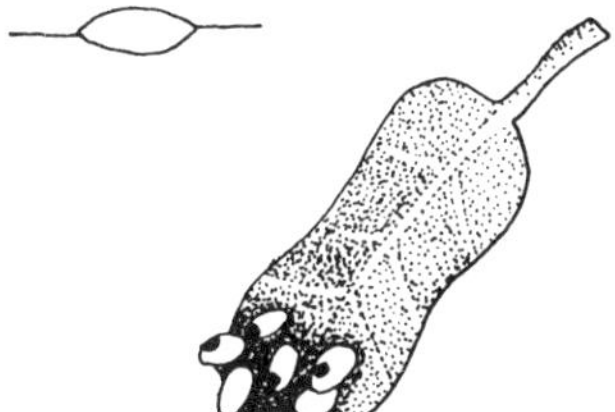

D. Bisexual generation galls of *Neuroterus saltatorius*

Fig. 121

their first appearance in May and can be found through September. In the early days of autumn the galls begin to drop by the thousands. These galls get their common name from the actions of the larvas once the galls have fallen from their host plants. The galls actually flip about on the ground as a result of the periodic, rapid movements of the larvas inside. The springing action lifts the galls up to 1 cm or more off the ground. Galls collected in September were still hopping in November, after being refrigerated to simulate natural conditions. All of this movement probably results in the downward direction of the galls into the leaf litter of the woodland. An extremely high percentage of the galls is parasitized. Those galls that remain on the leaves into the late fall and bear single tiny exit holes are usually parasitized. Those that escape their enemies contain prepupas by October, releasing parthenogenetic females over a 20-day span between March and the first of April. The bisexual generation develops within monothalamous, integral leaf blisters. These galls measure about 2 mm long by 1 mm wide. When fresh, these galls are green, but turn brown as the galls age. Researchers have found that more than half of the adults emerge between 4:45 a.m. and 6:30 a.m., with the remainder exiting throughout the morning. All of the females seem to leave by 1:30 p.m. In addition, Rosenthal and Koehler discovered during their 1971 studies that none of the adults lived more than 24 hours after emergence.

The urchin gall wasp, *Antron echinus* (see back cover), induces spectacular thick-spined, monothalamous, detachable, hard leaf galls on blue and scrub oaks. These summer galls, which bear a strong resemblance to sea urchins, are dark red to purple and measure up to 1 cm in diameter. The blunt spines are 3-4 mm long, and radiate outwardly over the entire surface of the rounded galls. The spines often have a slight bend and/or split tips. While most of the galls are not glossy, some have a slight varnish to them. These unisexual galls occur singly or in large numbers per leaf. By fall most of the galls have dropped from the undersides

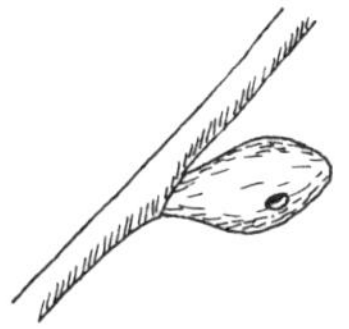

A. Bisexual generation galls of *Antron echinus*

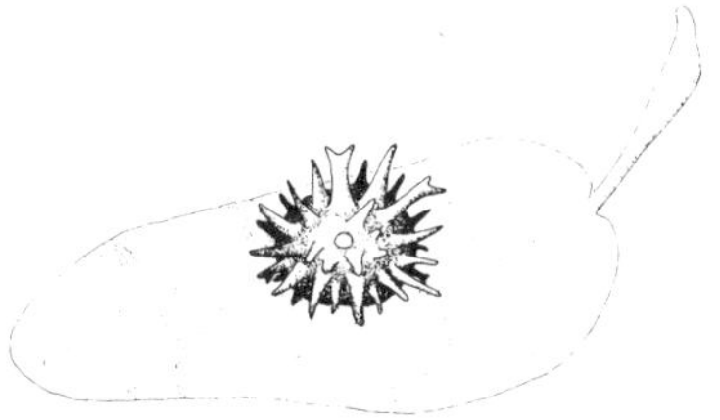

B. Unisexual generation galls of *Antron echinus*

of the leaves or have fallen with the leaves. Adults are found in the galls in November and emerge soon thereafter. Parthenogenetic females oviposit in the buds of blue and scrub oaks. The bisexual gall that results is succulent, thin walled, hollow, green, and globular in shape. From these rarely noticed 6 mm-long galls emerge the males and females. Like so many other cynipid galls, it is the unisexual summer galls that are the most spectacular in design and color. A second color form of this unisexual gall occurs in Sonoma County. This form is dominated by pink tones on the main gall body with white-tipped spines. If one collects, crushes, and boils the red form of these galls (hopefully, after the adults have emerged), one will get a deep wine-red color, which may be used in dyeing. If kept in a closed container, these galls yield a fragrance resembling raspberry.

The spined turban gall wasp, *Antron douglasii* (see back cover), induces flat-topped, monothalamous, detachable, spiny

C. Bisexual generation galls of *Antron douglasii*

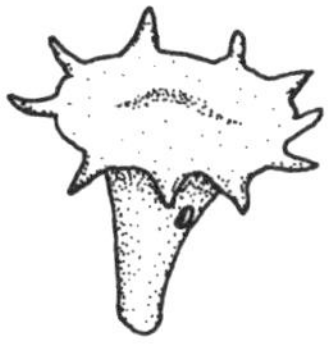

D. Unisexual generation galls of *Antron douglasii*

Fig. 123

galls on the leaves of blue, valley, and scrub oaks. The crownlike galls have narrow basal attachments measuring about 2 mm wide. From the narrow base the sides of the gall flare up and outwardly to the flat-topped crown, which bears lateral, stiff spines. At the broadest point the gall measures about 15 mm wide and reaches a height of about 10 mm. Often a slight dome or umbo develops at the center of the nonglossy, lavender to purple galls. The galls become glossy only after the smoky-grey bloom is rubbed off the surface. The larval chamber is located near the top of the gall, and an empty cavity develops just below the larval chamber near the base. The function of this cavity is unknown. Adults emerge in January and February to oviposit in the buds of the same host trees. In the spring less conspicuous bisexual galls develop. These reach a length of about 7 mm. Like the galls of *A. echinus* (bisexual generation), the bisexual galls of this species are succulent, green, but with longitudinal stripes and tubercles scattered over the surface, and a pointed apex. The males and females emerge in May and oviposit in the half-grown leaves. By late June the unisexual generation crown galls are well developed. Kinsey described several varieties of both generations. On the same trees that support the spined turban galls, there occur flat crown galls with more elongated, lateral spines, but without a long, stalklike neck. They rise to about 3 mm in height. These galls are often referred to the same species of wasp. Since a single tree can support thousands of the flattened crown galls, they are either a variety of *Antron douglasii* or an undescribed species. Certainly, their unique form is no fluke, but is a matter of regularity.

The club gall wasp, *Antron clavula,* induces detachable monothalamous, pubescent, club-shaped leaf galls on valley oak in the summer months. The galls normally occur singly on the veins on the undersides of the leaves, and measure 7-8 mm high by 2 mm in diameter. The cylindrical galls bear a bulge 3/4 of the way toward the tips. The bulge represents the larval chamber. There is usually a basal collar on most of these galls. When

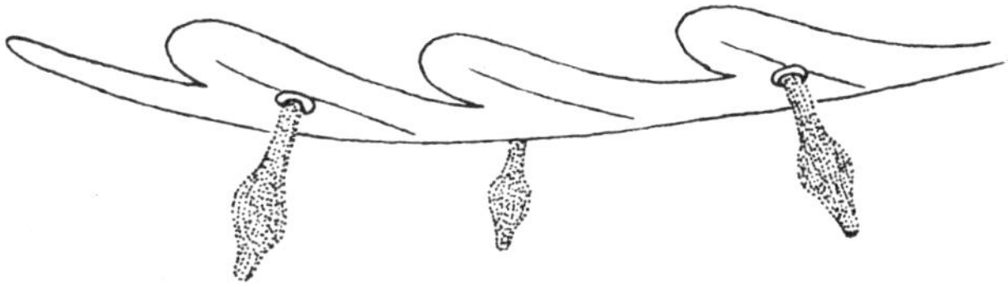

A. Cynipid wasp gall, *Antron clavula*

fresh, the galls are light beige with pink apices. Later the galls turn brown. A single gall on a lateral vein causes sufficient disruption in the flow of nutrients to the outer edges of the leaf that browning and death of the tissues result. Adults emerge from the swollen chambers in the spring. No alternate generation is known at this time.

The dunce cap gall wasp, *Andricus atrimentus,* (see back cover) induces detachable, white-creamy, red-streaked, cone-shaped galls on both sides of the leaves of blue and scrub oaks. From the flared bases, which touch the leaf along the edges of the gall, the sides rise to a pointed tip. These monothalamous summer galls are about 4 mm high and wide. The flared base is either smooth or toothed along the edge. These hollow, thin-walled galls are covered by a sparse coating of hairs. The larval chamber is located at the base of the gall. In late winter the adult breaks

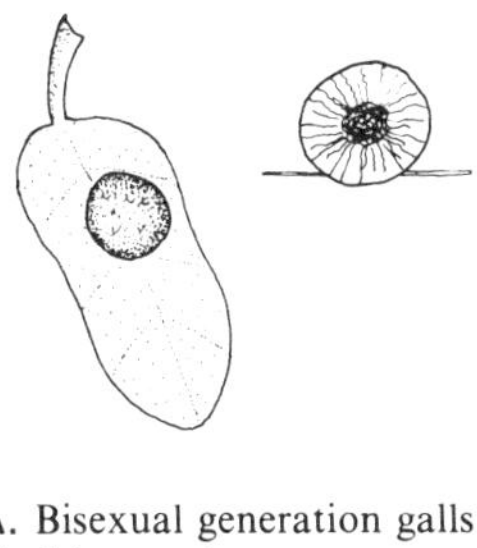

A. Bisexual generation galls of *Andricus atrimentus*

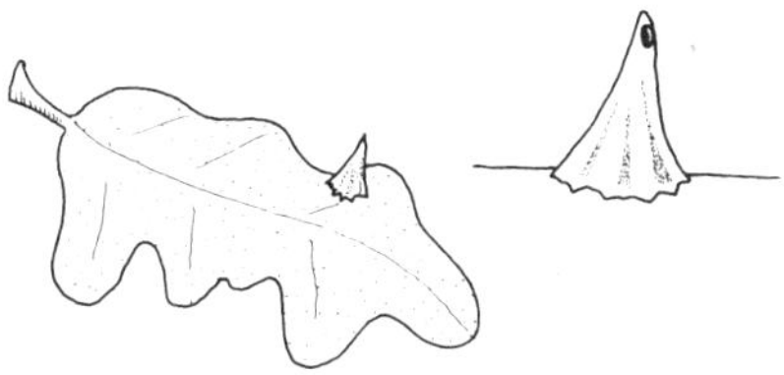

B. Unisexual generation galls of *Andricus atrimentus*

Fig. 125

out of the chamber into the cavity above, from which it emerges through a hole near the apex. These unisexual generation females oviposit in the leaf buds. By mid-March the bisexual generation galls are fully developed. These galls are characterized by their pink tissue-paper texture and the centrally located, blue-black larval chamber supported by numerous black filaments. Males and females emerge from these galls in April within a two-week period. Within 90 days after oviposition, the leaf galls of the unisexual generation are well developed.

The wasp *Dros pedicellatum* induces thin, horsehairlike, stalked, monothalamous, pointed leaf galls along the margins on blue, valley, scrub, and Oregon oaks. This early spring gall is further characterized by a noticeable bulge near the apex. This marks the location of the larval chamber. The galls appear as extensions of the leaf veins. They measure up to 2 cm long by about 2 mm in diameter at the bulge. The stalk beneath the chamber is less than 1 mm thick. When fresh the galls are brown, glossy, and hairless. These smooth galls are fully developed by April, with males and females emerging between April and June. Galls can still be found on the leaves in autumn. This would appear to be the bisexual generation of an as yet undescribed unisexual summer/fall generation. Further research may reveal the gall of the alternate generation, should one exist.

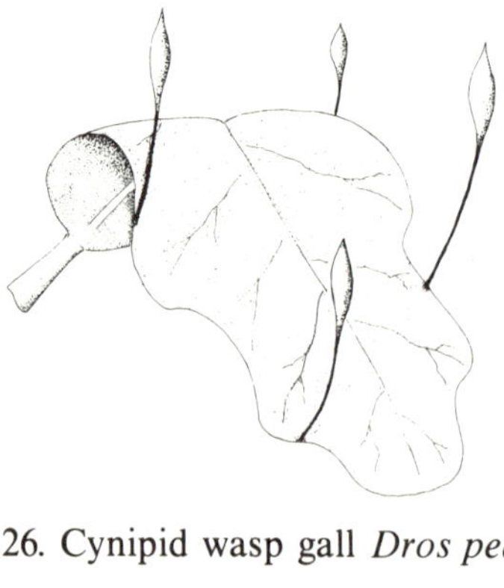

Fig. 126. Cynipid wasp gall *Dros pedicellatum*

Canyon Live Oak Galls

Canyon live oak and huckleberry oak are interesting in that the galls they support generally lack the brilliance of color and artistry of design found in the galls of white oaks. Of particular interest with this group of oaks is the occurrence of the same eriophyid mite found on the leaves of the black oaks. The other galls found on these live oaks are restricted to them and are caused by wasps.

Stem Galls. Most of the stem galls found in this group are rather ordinary in shape with an exception or two. Some of those referred to as stem galls are actually bud galls, which for the uninitiated may be too subtle and unimportant a distinction to make.

The split twig gall wasp, *Dryocosmus asymmetricus,* induces integral, polythalamous stem galls that rupture and split,

Fig. 127. Split twig gall, *Dryocosmus asymmetricus*

exposing a nestlike cluster of larval chambers. The nest of larval chambers resembles a clutch of brown elliptical eggs. The chambers are not separated by a thick mass of cell tissue as with other polythalamous galls. When fresh the galls appear as mere swellings of the stems. As they age during the summer months, the outer walls of the galls rupture widely, creating large openings into the interior of the galls. The galls reach 5 cm in length by 2 cm in width. The gall is normally asymmetrical with the larval chambers located in compact groups on one side of the central core of the gall. This might be the reason for the little damage done to leaves and branches beyond the gall. The flow of nutrients may be able to continue through unaffected tissue on one side of the gall, while the energy on the other side is directed into the support of the larvas. One might consider it an advantage to the wasps to have to chew their way out only through the walls of the chambers and not the entire gall. However, one might also consider the distinct disadvantage of increased susceptibility to parasitic attack or predation. Galls collected from Mt. Diablo in February had already been vacated by the adults.

The gall wasp *Heteroecus dasydactyli* induces detachable, globular, monothalamous, woolly stem galls on canyon live oaks in the summer. The light yellow-beige, dense, woolly mass measures up to 3 cm in diameter. The long hairs rub off easily in mats to expose the hard, solid, glossy gall beneath. The gall measures about 10 mm in diameter by 25 mm in length. The galls occur singly or in tightly clustered groups. The naked gall bears a ring or collar at the base. The 3 mm-thick neck of the gall expands upward into the bulging main body. The bulge tapers toward the apex, which is often recurved. While some galls are straight, others are drastically curved and bent. The surface of the gall is beige and covered with scattered tubercles and furrows near the base and apex. Fresh galls are light green to straw yellow. While the larval chamber is located centrally, the adults emerge near the base of the gall in the spring.

The wasp *Heteroecus melanoderma* induces elongate, monothalamous, woolly, detachable stem galls on both canyon live

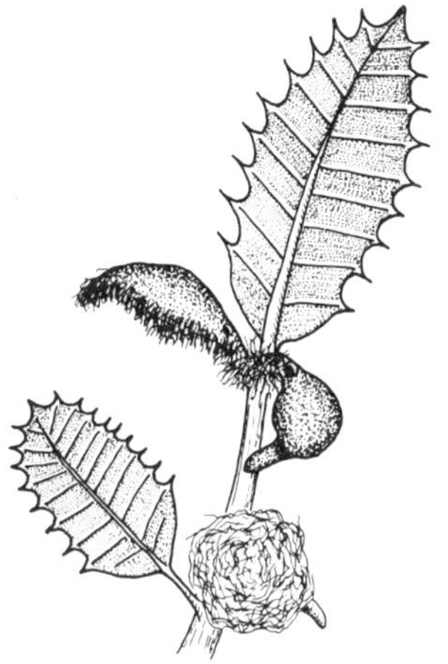

A. Unisexual generation gall of *Heteroecus dasydactyli*

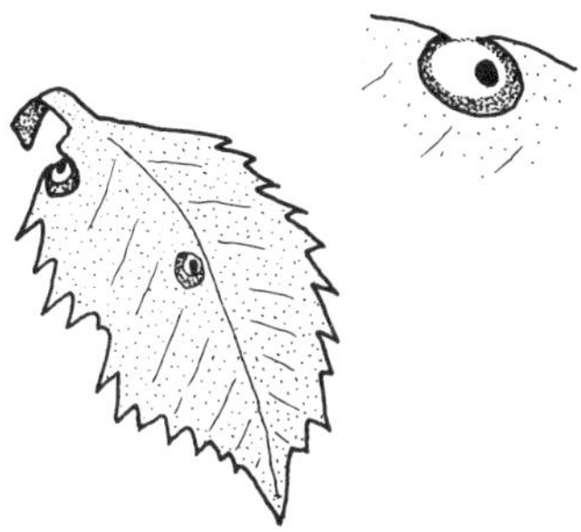

B. Bisexual generation gall of *Heteroecus dasydactyli.* Monothalamous, round to oval swellings protruding from both surfaces of leaf, 2 mm diameter.

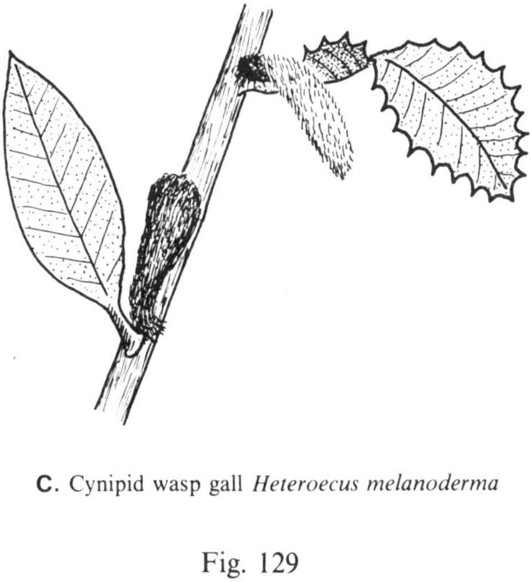

C. Cynipid wasp gall *Heteroecus melanoderma*

Fig. 129

oak and huckleberry oak in the summer. The short, yellow-beige hairs do not rub off easily in contrast to the above-named species. Instead, the matted hairs of *H. melanoderma*'s gall rub off only with persistent scratching. The hardened gall beneath has a nonglossy, mealy-granular appearance. The gall lacks a collar at the base, from which the sides rise and flare out to form the club shape typical of this species. The galls measure up to 20 mm long by 6 mm wide at the tips. The larval chamber is centrally located. Adults emerge through the neck near the base of the gall. While these gall wasps are common on canyon live oaks in the Sierra foothills, they are not so common on hosts along the central Coast Ranges. Unlike the galls of *H. aasydactyli*, whose fluffy hairs are often dislodged in rainstorms, the coating of hair on *H. melanoderma* galls persists.

The beaked spindle gall wasp, *Heteroecus pacificus,* induces spindle-shaped, monothalamous, hairless, detachable summer galls on the stems of both oaks. These unisexual galls are often elongate, measuring 30 mm long by 8 mm in diameter. Other specimens are squat and more robust. Some forms resemble upside-down pears with recurved beaks in place of the centrally depressed core. When fresh in July and August, the galls are normally succulent, solid, mottled green and red or various shades of green or, in some forms, beige with a slight waxy bloom. Fresh galls are usually smooth and glossy, while older galls have either wrinkled or cracked surfaces. The elliptical larval chamber is centrally located in the solid mass of tissue. The galls occur singly or in tight groups around the nodes of the stems.

Robert Lyon found that inquilinous insects were usually associated with the robust forms of these galls rather than with the slender forms. Lyons has suggested that the inquilines may have some influence on the shape of the galls. Parthenogenetic females emerge through holes situated near the base of the galls in April and May. The sexual generation gall develops soon after oviposition on the undersides of the leaves. These conical, thorn-shaped, monothalamous galls are 5-6 mm long by 2-3 mm in

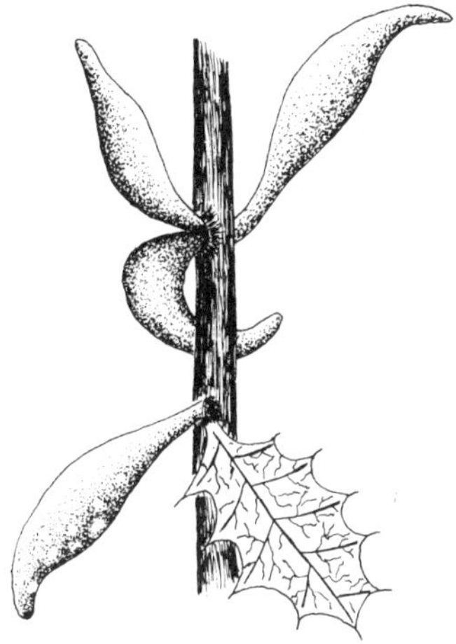

A. Unisexual generation galls of *Heteroecus pacificus*

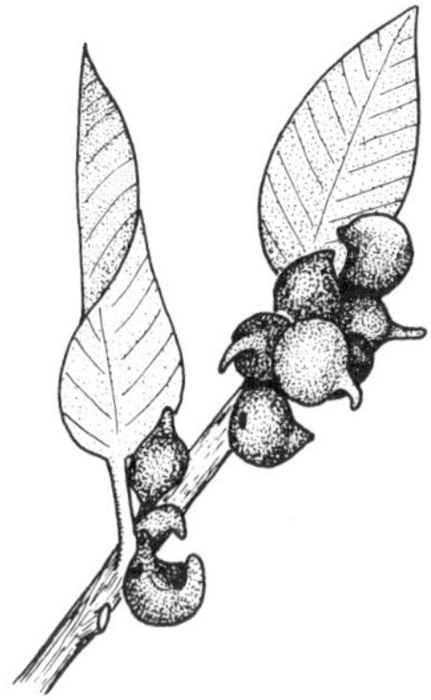

B. A second variety of the unisexual galls of *Heteroecus pacificus*

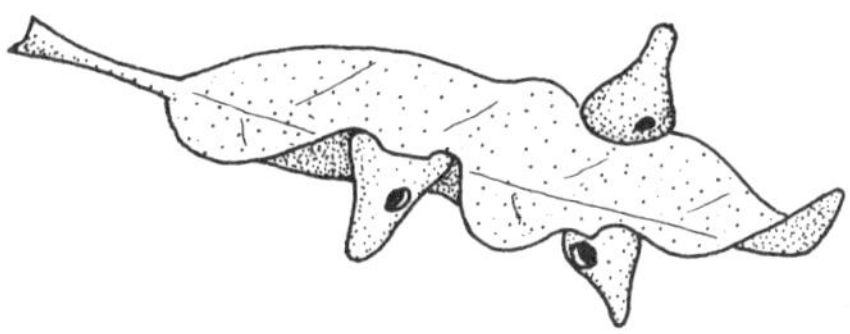

C. Bisexual generation galls of *Heteroecus pacificus*

Fig. 131

diameter. A dark depression 1-2 mm deep is usually noticeable on the upper surface of the leaves above each gall. These galls occur singly or in groups. The galls are not easily removed without damage to the leaf tissue. Where several galls develop in close proximity, the supporting leaf may be severely malformed. When fresh, the galls are red, but turn brown with age. Even though the infested tree may support hundreds of the unisexual galls, it is difficult to locate the bisexual galls. Males and females emerge from the thorn-shaped galls in June. It is interesting to note that Lyon discovered that out of certain clusters of these galls emerged only males, while other clusters produced only females.

The mushroom gall wasp, *Heteroecus sanctae-clarae,* induces monothalamous, detachable bud galls that closely resemble mushrooms or, with a little imagination, Russian church steeples. The two-story galls are characterized by round basal sections that are capped with pyramidal, nippled upper sections or tops. The upper half of the galls normally rises to a recurved nipple at the apex. These sections are nonglossy, smooth, or faintly tubercled and furrowed at the apex. The upper half of the gall is easily removed from the bottom section by a twisting motion. This action reveals an erect, vertical column that is attached to the base and snugly fits into the upper section. It is

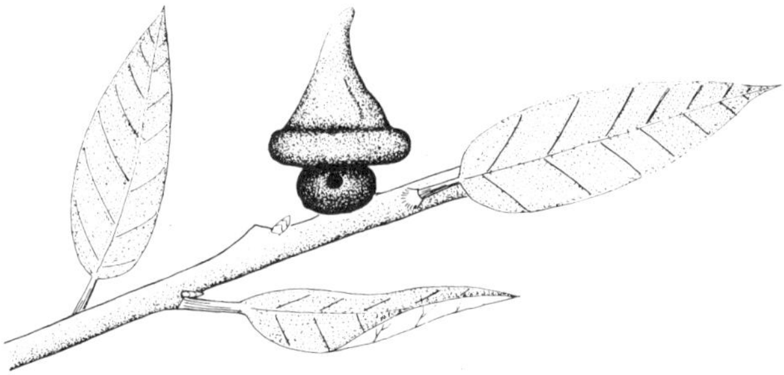

Fig. 132. Mushroom gall, *Heteroecus sanctae-clarae*

within this vertical column that the elliptical larval chamber is located. The entire gall measures up to 1.5 cm high and wide. Generally, the two segments of the gall are equal in size, but occasionally the upper half is much larger than the bottom. The galls normally stand erect on the stems, and occur singly on both species of oaks. Adults emerge from the base in April and May. Gall development reaches its peak by July in most areas. No alternate generation is known.

The wasp *Disholcaspis chrysolepidis* induces hard, detachable, humped, polythalamous galls that burst out of the stems in rows on both oaks. These early summer galls are narrowly attached

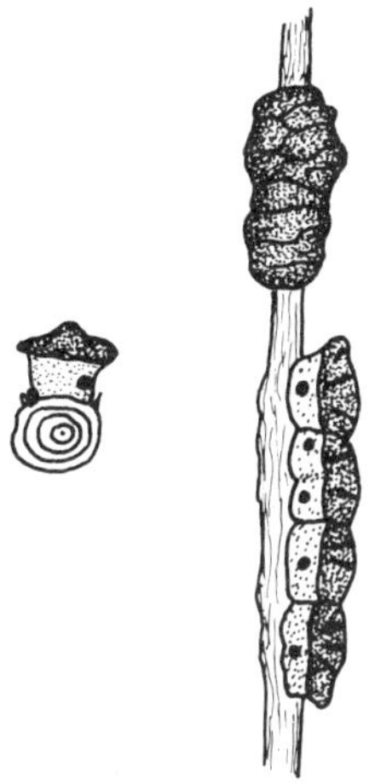

Fig. 133. Cynipid wasp gall *Disholcaspis chrysolepidis*

and thus easily removed. They occur singly, but more commonly appear in clustered rows. Fresh specimens are dark rusty brown on top with light brown sides. Often young galls have greenish-tan colors along the sides. The convex tops have a cracked, bread-crust texture, while the sides are relatively smooth. A single gall can measure up to 10 mm in diameter, but the clustered rows of galls may reach 3 cm in length. Each gall is characterized by an elongate ridge that parallels the stem. From the crest of this ridge, the gall flares down and out to the overhanging edge. The sides of the gall narrow down and in towards the basal attachment. The stems supporting the galls are not noticeably altered except for the split where the galls erupted. An individual gall rarely supports more than two larvas, each in its own chamber. When fresh, these galls, like other cynipid galls, release a sweet phloem exudate, which attracts ants, bees, and other insects. Parthenogenetic females emerge through the sides in midwinter. An alternate generation is not yet known.

The wasp *Andricus spectabilis* induces integral, nondetachable, abruptly swollen, polythalamous, large stem galls on canyon live oak. The bark-colored swelling measures up to 6 cm long by 3 cm in diameter. The texture is much like that of the bark, i.e., rough with faint cracks. Damage to the branches and leaves beyond the galls has not been observed. Gall development begins in late spring and reaches its peak by midsummer. Females emerge during the following spring. Little else is known about this species.

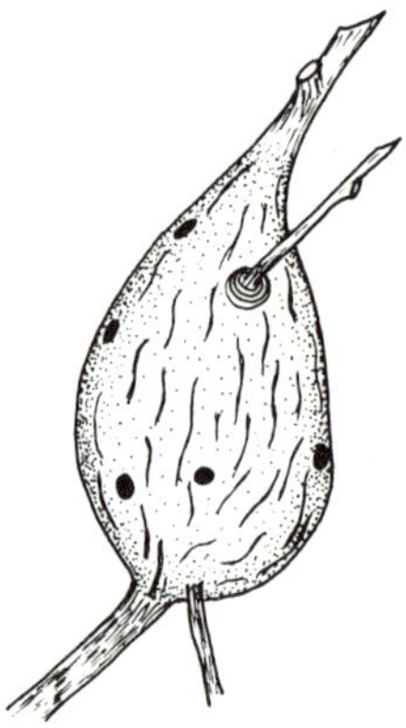

Fig. 134. Cynipid wasp gall *Andricus spectabilis*

The wasp *Disholcaspis truckeensis* induces lightweight, detachable, polythalamous, smooth, potato-shaped galls on the stems of both oaks. At maturity the tan-colored, slightly glossy gall measures about 6 cm long by 3 cm in diameter. This orbicular summer gall is characterized by an extremely thin outer skin, which cracks with age. These cracks often originate at wasp exit holes and exhibit a slight peeling quality. Elsewhere patches of the skin fall off, exposing the soft, pulpy inner tissue of the gall. Normally no more than a few galls occur on any one branch. Little, if any, damage results to the host trees as a result of the activities of this gall wasp. Parthenogenetic females emerge in January and February.

The kidney stem gall wasp, *Andricus reniformis,* induces detachable, polythalamous, kidney-shaped summer galls on huckleberry oaks. When fresh, the galls are mottled red, yellow, and green; with age, the galls become tan and light yellow. Galls measure up to 3 cm long by 1.5 cm in diameter and are solid. The larval chambers are arranged radially around a central focal point or core that runs the length of the galls. The gall wasps do not emerge from the galls until two or three years after gall formation. Specimens collected in 1973 still contained preadult forms (actually prepupas) in the spring of 1976. Adults emerged within six weeks after reexamination of the galls. This species is extremely common from the Yuba Pass south to the Sonora Pass in the central Sierra.

A. Cynipid wasp gall *Disholcaspis truckeensis*

B. Kidney stem gall, *Andricus reniformis*

Fig. 135

The paper gall wasp, *Trichoteras vaccinifoliae,* induces round, hollow, thin-walled, monothalamous, detachable summer galls on the stems of huckleberry oaks. The galls are characterized by the thin, fleshy, mottled red, yellow, and green outer wall and the centrally suspended larval chamber supported by numerous radiating fibers that connect it with the walls. These lightweight galls measure up to 3 cm in diameter. When fresh in early to midsummer, these galls are somewhat succulent and easily damaged by improper handling. Later the galls become dry, extremely brittle, paper thin, and turn light brown in color. By September most of the galls have reached the dry, wrinkled stage. The surface of these galls bears a slight sheen at all times. Although seemingly smooth, the surface of these galls is characterized by slight umbos or lumps. Most of the galls tend to develop on older sections of the branches rather than on new spring growth. Fresh specimens examined in the field in August were found to contain mostly last-instar larvas. Old, dried galls examined at the same time contained prepupas and were subsequently placed under refrigeration to simulate natural winter low temperatures. These specimens were removed from the cooler in late April of the following year, and yielded adults

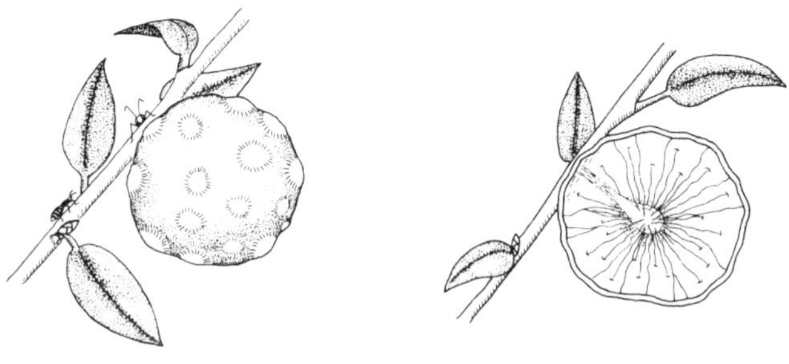

A. Paper gall, *Trichoteras vaccinifoliae*

B. Cross-section reveals the fibers that deliver nutrients to the central larval chamber

Fig. 136

within three weeks. Adult emergence is probably timed with snow melt and warming temperatures. This wasp has also been reported to gall canyon live oaks, but I have not yet found this to be the case. However, the species is extremely abundant throughout the Sierra and wherever huckleberry oaks thrive.

Leaf Galls. The leaf galls on this group of oaks are interesting in that most of them have not been identified, leaving only a few that are well known. As with the leaf galls of the black oak group, the galls of this group are not as spectacular as those of white oaks.

The gall wasp *Andricus lasius* induces round, hairy, polythalamous, detachable, midrib vein galls on both sides of the leaves of canyon live oaks. A similar gall occurs on huckleberry oaks. The key characteristic of these galls is the dense mat of golden-brown hairs that turn dingy grey with age. The hairs are about 3 mm long. The entire gall measures up to 10 mm in diameter with the hairs included. The galls usually occur singly on the midrib vein, but are also known to develop on the lateral veins and at the base of the leaf and petiole. The larval chambers, which normally number fewer than six, are radially arranged just above the basal point of attachment. Females emerge from January to early March. Although commonly found within the range of the host plant, the life habits and cycle remain largely unknown.

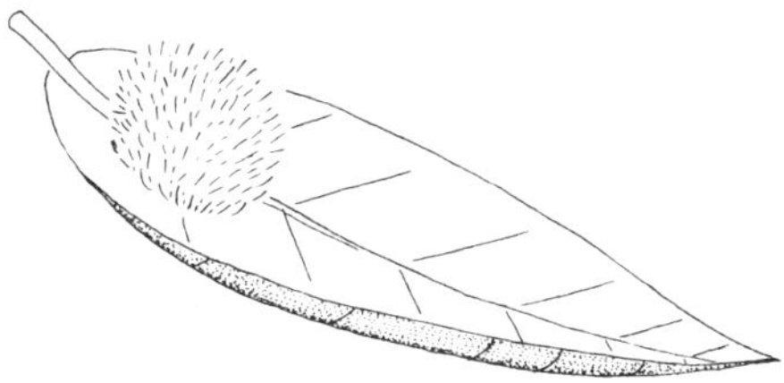

Fig. 137. Cynipid wasp gall *Andricus lasius*

The wasp *Paracraspis guadaloupensis* induces convex, flat, disc-shaped, monothalamous galls on the undersides of the leaves of canyon live oaks. These spring, beige disc galls occur singly or several per leaf, but rarely are they found touching in group fashion. Usually only one gall occurs per lateral vein. The galls measure up to 6 mm in diameter. The central larval chamber is nearly half as wide as the entire gall. The galls are extremely thin, easily damaged and removed. Gall development begins prior to May on last year's leaves. Adults emerge from late winter to early spring. The exit holes of the females is slightly off center. No visible damage results from the activities of this wasp or from the growth of the gall. This species appears not to be abundant in any one area I have explored.

The wasp *Paracraspis patelloides* induces saucer-shaped, detachable, monothalamous, deeply depressed, flat-rimmed galls on the lateral veins of last year's leaves on canyon live oaks. In late April in the central Sierra the young, fleshy galls are rather flat to slightly convex with prominent central depressions that form deep bowls. The disc-shaped galls are about 4 mm at this time. The color of these galls is a faint pea-green to ivory with pinkish-red margins. By summer the sides of the galls rise about 5 mm above the surfaces of the leaves in a perpendicular fashion. The 2 mm-wide, flared rim becomes prominent by this time as it surrounds the deep depression of the inner cup. The entire gall measures about 10-12 mm in diameter. The sides of the galls are usually ridged and nonglossy. The color of mature galls is dominated by beige tones with hints of pink and red along the margins of the rim. Close inspection reveals small dark spots over the entire surface of the gall. The larval chamber is located centrally at the base of the depression. A hollow cavity often develops beneath the larval chamber. Adults are found in the galls in November and December. Emergence takes place sometime thereafter through holes in the center of the galls.

Cynipid research appears to be a wide-open field for the curious and certainly the source of innumerable studies and

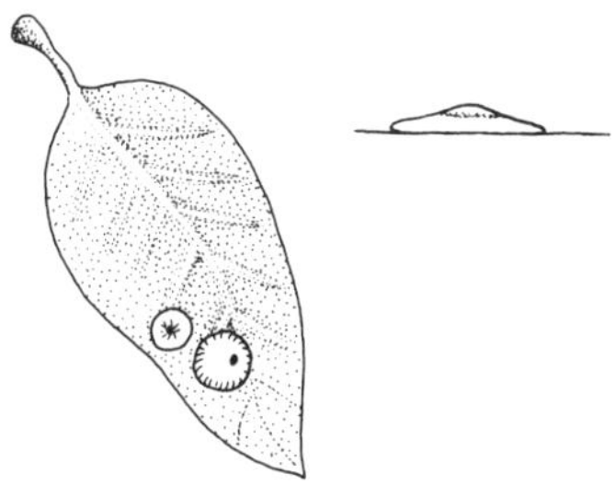

P. guadaloupensis

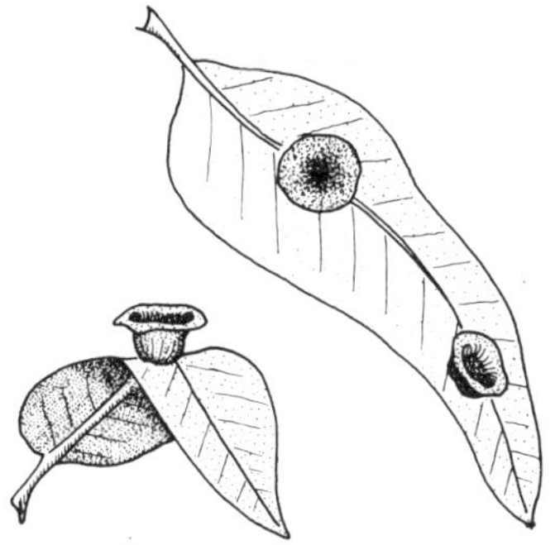

P. patelloides

master's and doctoral theses. So much has yet to be learned before we fully comprehend the levels of sophistication in adaptation among these insects.

The eriophyid mite *Eriophyes mackiei* induces rust-colored erineum patches on the undersides of both huckleberry and canyon live oaks, as well as on members of the black oaks. These erineum galls appear as hairy depressions on the undersides of the leaves with corresponding bumps on the dorsal sides. The

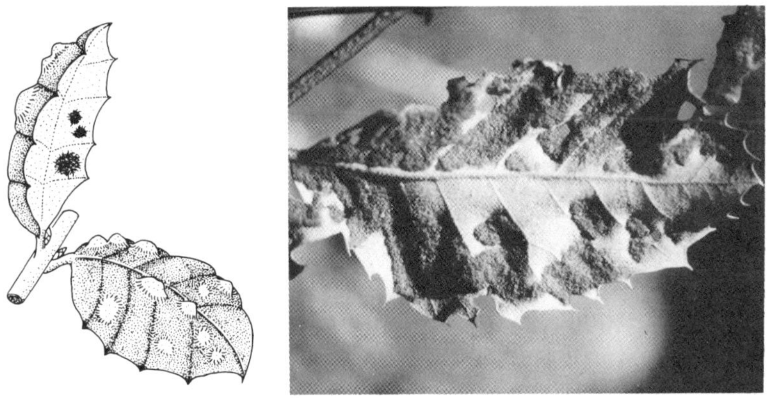

Erineum gall, *Eriohyes mackiei*

galls on huckleberry oak are often only 2 mm in diameter, and several dozen galls occur per leaf. The erineum patches on canyon live oak often encompass the entire undersurface of the leaves. It seems a bit strange that in a world with such prevalent host specificity, a gall organism such as this mite would also be found on other oaks outside of this group. Of course, this may point to a close genetic relationship between the black and canyon live oaks or to the tenacity of the mite.

Pine Galls

There are at least 20 species of pine trees spread over the state from the coastal areas to the high Sierra and upper desert regions. Thirteen of these species occur naturally out of the boundaries of California and most likely support the same gall organisms. Seven species of pines are restricted to California. The height of many pines makes it difficult to observe galls except on young trees. Two different gall organisms, western gall rust and dwarf mistletoe, regularly attack the following pines in California: Bishop pine, *Pinus muricata;* digger pine, *P. sabiniana;* whitebark pine, *P. albicaulis,* knobcone pine, *P. attenuata;* Coulter pine, *P. coulteri;* pinyon pine, *P. monophylla;* limber pine, *P. flexilis;* Jeffrey pine, *P. jeffreyi;* sugar pine, *P. lambertiana;* western white pine, *P. monticola;* ponderosa pine, *P. ponderosa;* lodgepole pine, *P. murrayana;* Monterey pine, *P. radiata.* See Munz for detailed range descriptions of these species.

The western gall rust, *Peridermium harknessii,* induces large, abruptly swollen, globular, integral stem galls on knobcone, Coulter, Jeffrey, ponderosa, digger, and Monterey pines. Characterized by its large size, which ranges up to 35 cm in diameter, rough texture, and orange spores produced in the spring, this gall frequently is responsible for the death of the branches beyond the gall. In addition to interrupting the flow of essential nutrients, the fissured gall provides inroads into the sensitive tissue of the trees for other pathogenic organisms. The

A. Western gall rust, *Peridermium harknessii.* Photo shows galled timbers used to support lodge roof in Yellowstone National Park.

bright orange spores show through the deep fissures and, in addition to the large size, make the galls quite noticeable. After sporulation, white, filamentous material fills the fissures. Unlike some other rusts, which have an intermediate host, western gall rust goes from pine to pine. Monterey pines that are planted ornamentally can be severely damaged by this fungus, particularly where surface roots are exposed and wounded through lawn mowing. Proper pruning prior to spring will help in controlling some of the damage.

The dwarf mistletoe, *Arceuthobium campylopodum* and *A. americanum,* induces elliptical, integral stem swellings on nearly

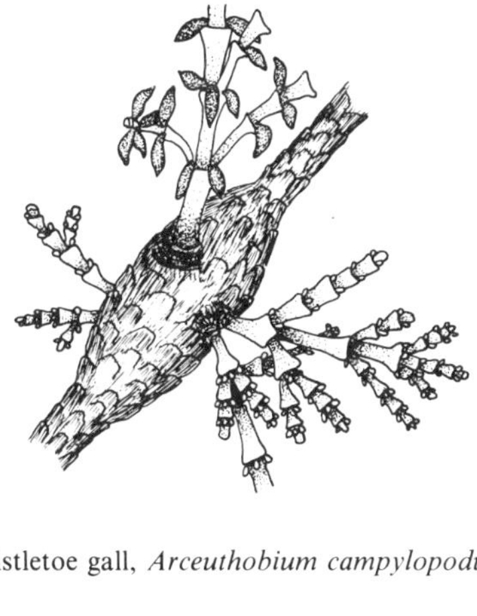

B. Mistletoe gall, *Arceuthobium campylopodum*

Fig. 141

all of the pines aforementioned. The galls measure up to 15 cm long by 8 cm in diameter. The swellings are noticeable beneath a burst of succulent, short, bright yellow-orange, nearly leafless mistletoe stems. In addition, the stems of the mistletoe are obviously jointed. Often the elliptical galls are hidden beneath numerous shoots of mistletoe and pine in a witches' broom fashion. Galls spread as the sticky seeds of the mistletoe land on uninfected branches, germinate, and incite the formation of the galls through the feeding actions of the parasite. *Arceuthobium campylopodum* is a serious pathogen on Jeffrey pine, while *A. americanum* commonly attacks lodgepole pine.

Spruce Galls

There are three native species of spruce in California, Sitka spruce, *Picea sitchensis;* weeping spruce, *P. breweriana;* and Engelmann spruce, *P. engelmannii.* Sitka spruce grows along the coast below 365 m from Mendocino County north into Alaska. Engelmann spruce is found at the 1300-m level in Shasta and Siskiyou counties and north into British Columbia. Weeping spruce grows from 1300-2300 m in Del Norte, Trinity, and Siskiyou counties and into southwestern Oregon. Each of these species of spruce is galled by the same organism.

The Cooley spruce gall aphid, *Adelges cooleyi,* induces integral, elongate, needled stem galls near the tips of the branches. When fresh in the spring, these galls are green, fleshy, and covered with normal-sized needles. With age, they turn brown, generally lose their needles, and appear more like cones with spreading and open bracts. These galls normally reach up to 4 cm long by 1-2 cm in diameter. Some larger specimens have been recorded. The galls are characterized by liplike slits that develop above the base of each needle.

This gall insect, which is not technically an aphid, overwinters as an immature female on the host tree. It matures in the spring as as adult stem mother and lays several hundred eggs on the terminal shoots of spruce trees. Later, nymphs migrate to new

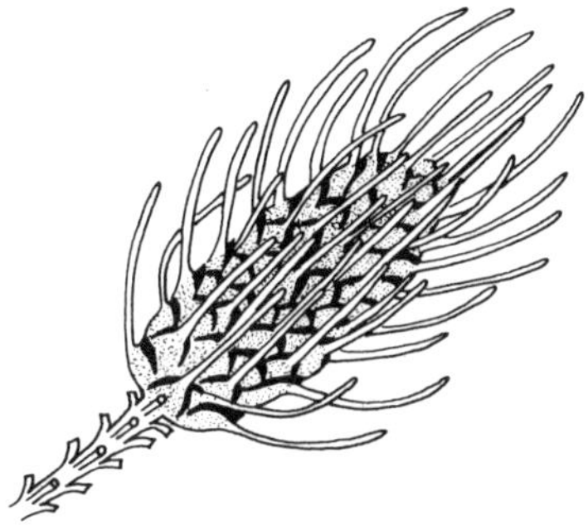

Fig. 143. Cooley spruce gall, *Adelges cooleyi*

spring growth and feed at the bases of the needles. Their actions and secretions force the plant tissues to swell around them, forming the gall. By midsummer, the liplike slits develop just above each needle around the entire gall. The adelgids escape through these slits and migrate to the tips of the needles where they turn into winged adult females. These females in turn migrate to either Douglas fir, *Pseudotsuga menziezii,* or to another spruce. Eggs laid on the needles of Douglas fir produce only "woolly aphids" and no galls. This adelgid is known to gall oriental spruce, *P. orientalis,* and Colorado blue spruce, *P. pungens,* in addition to our native spruces in this area. I have found these galls extremely common on the spruces of western Montana and Wyoming. In this area the adelgids are at the tips of the gall needles in August in the winged adult form. These galls have also been found with regularity in the Mt. Lassen area of California.

Walnut Galls

There are three species of walnuts in California, one of which is barely distinct from the others. These include the English walnut, *Juglans regia;* the California black walnut, *J. hindsii;* and the southern California walnut, *J. californica.* Attention will focus on *J. regia* and *J. hindsii* because of the apparent confusion over distinguishing characters of *J. californica.*

The walnut blister mite, *Eriophyes erineus,* induces large, feltlike masses of pale-yellow hair on the undersides of the leaflets of English walnut. The hair-lined depressions often measure up to 17 mm in diameter, but occasionally individual erineum galls coalesce to cover much larger sections of the leaflet. Convex lumps on the dorsal surface of the leaflets correspond to the erineum depressions beneath. It is thought that before leaves drop in the fall, the mites move into the leaf buds where they overwinter. These erineum gall mites have been found in 11 counties ranging in the north from Glenn and Butte counties to Riverside County in the south.

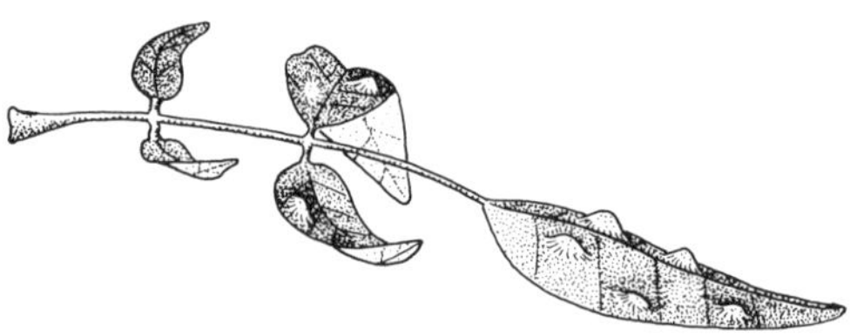

Fig. 144. Walnut erineum gall, *Eriophyes erineus*

The black walnut purse gall mite, *Eriophyes brachytarsus,* induces globular, rough-surfaced, green pouch galls on the dorsal surface of the leaflets of *Juglans hindsii.* The galls occur singly or in coalescing clumps. They measure from 3-8 mm in diameter, and readily stand above the surface of the leaflets on short pedicels. The undersurface of the leaflets is marked by depressed

Fig. 145. Walnut purse gall, *Eriophyes brachytarsus*

openings at the sites of the galls. Dozens of these galls can occur on each leaflet. In the spring only the stem mothers are present during the early stages of gall development. Later the galls harbor the reddish offspring. As the leaflets dry in the autumn, the galls turn from yellowish-green to black. These mite galls have been found from Shasta County in the north to San Diego County in the south.

Willow Galls

Munz recognizes over 30 species of willows, including several varieties, as occurring in California. To include detailed taxonomic and geographic data on all of these species is clearly beyond the scope of this work. Instead, attention will be focused on six species in keeping with E. L. Smith's 1968 work on the galls of willows. The discussion of the galls found on these species in no way implies a strict host restriction to the species mentioned, even though this is the case with some gall organisms. The galls found on willows are associated with three groups of organisms: tenthredinid sawflies, eriophyid mites, and cecidomyiid midges.

The six species of willows referred to in the gall descriptions include the following:

Narrow-leaf willow, *Salix exigua,* below 2400 m, deserts and east side of Sierra Nevada to British Columbia in wet places.

Arroyo willow, *Salix lasiolepis,* below 2100 m, many plant communities throughout California to Washington, along stream banks.

Lemon's willow, *Salix lemmonii,* 1500-3000 m, San Bernardino Mountains, Sierra Nevada, Tulare County north to Oregon, in wet places and coniferous forests.

Brewer's willow, *Salix breweri,* below 1370 m, inner Coast Ranges, San Benito to Lake and Colusa counties, gravel beds and ravines.

Scouler's willow, *Salix scouleriana,* below 3000 m, many plant communities, San Jacinto Mountains north through Sierra Range, from Monterey north to Alaska, wet places.

Geyer's willow, *Salix geyeriana,* 1500-3000 m, Mono County to British Columbia, Idaho, in high mountain areas.

Stem Galls. There are at least six common species of sawflies in the genus *Euura* that induce galls on California willows. Unlike other gall organisms, in which the larvas initiate gall development, the adults of this genus and the genus *Pontania* are responsible for gall formation. The galls of *Euura* sp. normally reach full size before the eggs hatch. Colleterial fluid injected into meristematic plant tissue by the adults programs the willows in the formation of the galls. In addition to the stem galls induced by these sawflies, two other stem galls are initiated by midges of the genus *Rhabdophaga.*

The sawfly *Euura exigua* induces linear, gradually tapering integral stem galls on *Salix exigua, S. melanopsis,* and *S. hindsiana.* These smooth-surfaced galls are either with or without minute hairs. At first, they are green, but later turn brown. They measure 2-7 cm long by .5-1.5 cm in diameter. The galls are usually lopsided with most of the development occurring on one side of the stem. The galls of this species resemble those of *E. lasiolepis*. Two or more galls may combine in development, because of close oviposition, to create what appears to be one large gall. The larva crawls to the top of the gall to cut an exit hole and plug it with frass prior to pupation. Each gall may contain more than one larva. Pupation occurs in the gall. Emergence of males and females occurs in the spring. New galls begin development shortly thereafter.

The sawfly *Euura geyerianae* induces bulbous, abruptly swollen integral stem galls on *Salix geyeriana*. Fresh galls are often covered with powdery white material. Mature galls are thick-walled, glossy green to mottled red-brown, and measure up to 2 cm long by 1.5 cm in diameter. The larva makes an exit hole prior to pupation in these polythalamous galls. Males and females emerge in the late spring. The species is abundant in scattered locations from Oregon through the high conifer forests

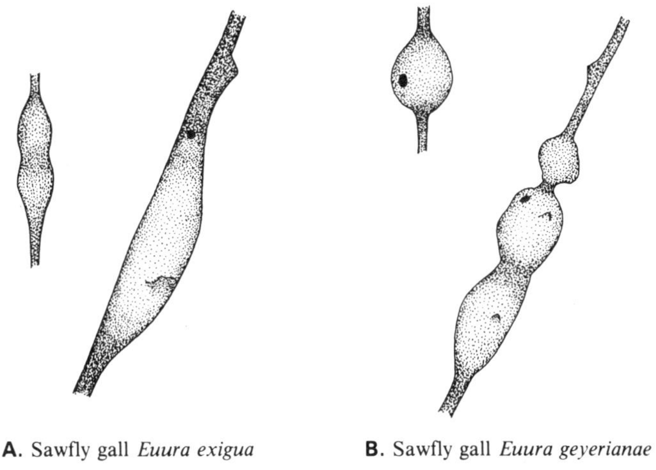

A. Sawfly gall *Euura exigua* **B.** Sawfly gall *Euura geyerianae*

Fig. 147

along the east side of the Sierra. Adults of this species and other members of the genus have been observed to feed on nectar, pollen, and the stamens of host willows.

The sawfly *Euura scoulerianae* is responsible for thick-walled, elongate, smooth integral stem galls on *Salix scouleriana.* The galls usually taper toward the top. They are occasionally pubescent, and are normally mottled green and red when fresh. Mature galls turn brown at first, then grey in later stages. The galls measure up to 4 cm long by 1.5 cm in diameter. Last instar larvas cut, usually in autumn, an exit hole near the top of the gall and plug it with frass before retreating to the base of the gall prior to pupation. Males and females emerge in May and June. Like other species of this genus, there is only one generation per year. More than 75% of the population of this species often succumbs to parasitism. Smith has found that pupation is influenced by the moisture content of the host branches in the spring. Branches broken or cut before the buds burst open usually do not yield adults.

The sawfly *Euura breweriae* induces linear, lopsided, smooth, glossy integral stem galls on *Salix breweri.* These thin -walled

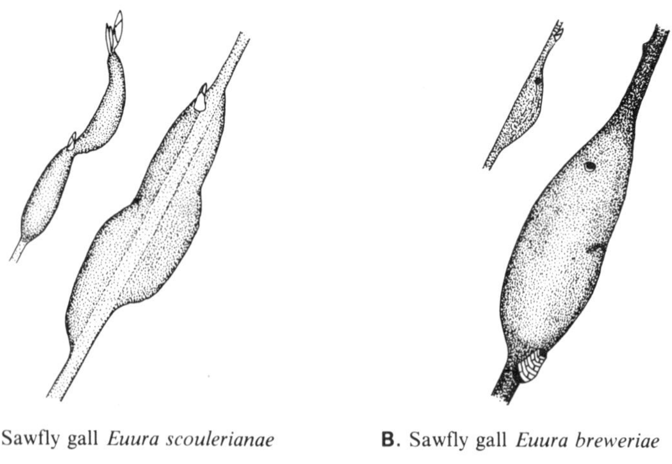

A. Sawfly gall *Euura scoulerianae* **B.** Sawfly gall *Euura breweriae*

Fig. 148

galls develop heavily on one side of the stem. When fresh, the gall color is often bright yellow-green, tinged with a hint of red. Young galls are pubescent. Mature galls reach 2-4 cm in length by 1.5 cm in diameter. These galls contain from 1-3 larvas each. The larvas of this species do not cut exit holes prior to diapause and pupation. Adults emerge in April and May along the inner Coast Ranges.

The sawfly *Euura lemmoniae* induces abruptly swollen, irregular, potato-shaped, somewhat knobby, integral stem galls on *Salix lemonii.* The galls measure up to 6 cm long by 1.5 cm in diameter. The massive swellings occur equally on both sides of the stems. Occasionally they occur mostly on one side of the stem. The surface of the gall is slightly striated, glossy to dull, and dark green to reddish brown when fresh. Normal-appearing buds

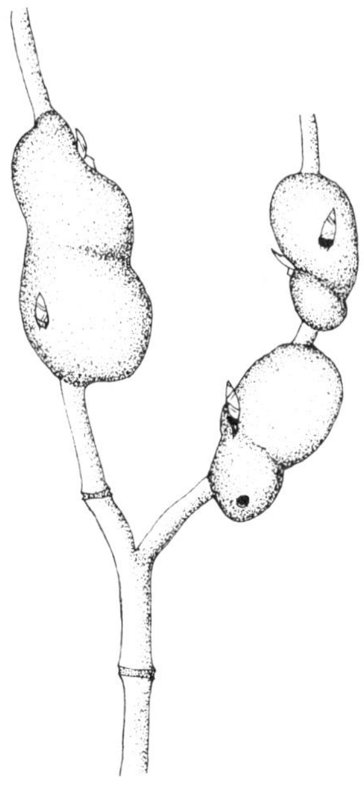

Fig. 149. Sawfly gall *Euura lemmoniae*

are frequently found on the sides of these galls. The larvas do not cut exit holes prior to pupation. Like other species of *Euura,* the larvas of *E. lemmoniae* have special defense mechanisms. They eject large quantities of a pungent fluid from abdominal glands, which are exposed by a lifting and waving motion of the posterior of the larvas when disturbed. Males and females emerge between late April and mid-June. This sawfly is the dominant member of the genus in high-altitude alpine meadows in the Sierra Nevada.

The sawfly *Euura lasiolepis* induces thick-walled, glossy, lopsided, integral stem galls on *Salix lasiolepis.* The galls found growing in the shade are normally yellow-green. Those exposed to the rays of the sun are red and purple. The galls occur singly on one side of the branch, with some galls developing opposite each other. These glabrous galls reach a length of about 7 cm with a width up to 2 cm. The tops of the galls gradually taper into the stems, while the bases of the galls are abruptly swollen. As these galls dry, they shrivel considerably. Each gall contains from 1-5 larvas, which do not cut exit holes prior to diapause and pupation. Smith has suggested that a disadvantage in precutting exit holes is the possibility of the flooding of larval chambers from winter rains, thus killing the insects.

While most other species of *Euura* overwinter in a prepupa state, the coastal race of *E. lasiolepis* feeds and grows at a slow rate until January. In late winter the prepupas metamorphose quickly in concert with the expansion of tree buds. In the Sierra this species pupates and emerges in May. The emergence of these insects is also timed with bud opening and shoot elongation among willows. These trees also give evidence of the existence of clones that are immune from the attacks of sawflies, while other clones are quite susceptible.

The willow rosette gall midge, *Rhabdophaga brassicoides,* induces detachable, open or closed rosette bud galls on several species of willows. The galls appear as compact clusters of leafy bracts, and superficially resemble cones. Many resemble cabbages with squat arrangements of leafy bracts. The cone-shaped form reaches 2 cm in height by 2 cm wide at the base. The

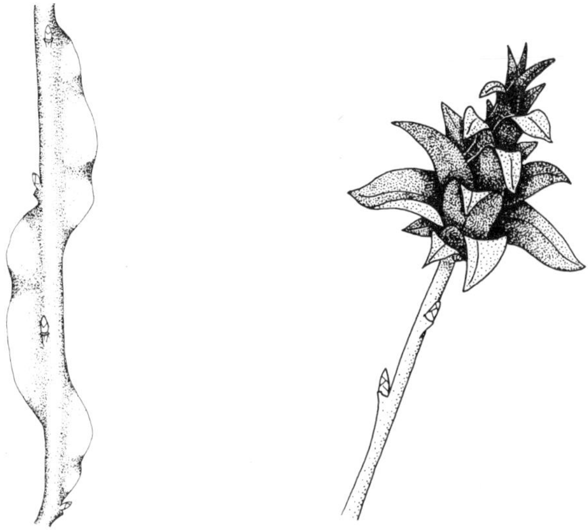

A. Sawfly gall *Euura lasiolepis*

B. Midge rosette gall, *Rhabdophaga brassicoides*

Fig. 151

leafy bracts extend away from the main gall body and measure 7-8 mm long by about 9 mm wide at the basal attachment. In the spring the galls are green and pliable. By fall they turn brown and become brittle. The larval chamber is located just above the base of the gall, and is surrounded by overlapping layers of leafy bud scales. In winter these willow rosette galls are easily seen at the tips of branches protruding from the snow at elevations of 1500-2100 m in the Sierra Nevada.

A second form of this gall lacks the narrowed, conelike tip of the former gall. These galls more closely resemble cabbages than cones. They measure up to 2.5 cm wide and high. The broad, leafy bracts reach 1.5 cm wide by 1 cm in length. These cabbagelike rosettes appear on willows at lower elevations. With further research they may prove to be a separate species, but for the time being they are considered as being induced by the same midge. Galls of both forms are often pubescent depending on the species of host willow. Adult midges emerge in the early spring at low elevations and later in higher altitudes. There appears to be only one generation per year.

The willow potato gall midge, *Rhabdophaga salicisbatatus,* induces irregular, knobby, potato-shaped, large, integral stem galls on a variety of willows. These polythalamous galls severely distort the general trend of the branches, causing them to twist and bend. Buds occasionally burst open from the sides of the galls, but fail to develop normally. Fresh galls are green and smooth. These robust galls measure up to 4 cm long by 2 cm in diameter. The dried, dark-brown galls found in the fall are extremely hard. Adults pupate at the surface of the galls in early spring. These galls seem to be fairly common at lower elevations. They cause damage to the branches and leaves beyond the galls.

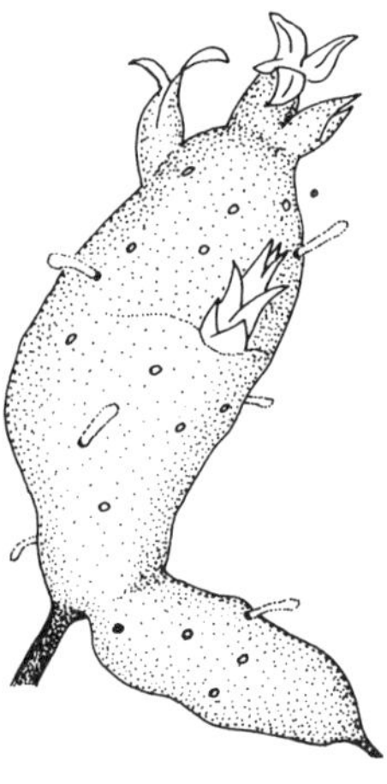

Fig. 152. Midge potato gall, *Rhabdophaga salicisbatatus*

Leaf Galls. The common leaf galls associated with willow leaves involve tenthredinid sawflies, cecidomyiid midges, and eriophyid mites. The sawflies are responsible for the majority of these galls, which are the largest and most colorful found on willows.

The willow apple gall sawfly, *Pontania pacifica,* induces bright green and red, round, monothalamous galls that protrude on both sides of the leaves of *Salix lasiolepis* and *S. lemmonii* and related species. The form of the galls varies from ovular to bean-shaped, but is more often found to be round and to resemble small apples. The galls vary in size from 7-11 mm at their widest point. They occur singly on either side of the midrib vein or in groups numbering as many as 14 per leaf. Tiny, wartlike scales called **lenticels** characterize these otherwise smooth galls. The galls may or may not possess fine hairs depending on the species of host willow. While many galls form equally on both surfaces of the leaves, others protrude mostly on the lower side. Galls exposed to direct sunlight are usually bright red on top and yellow-green below. Gall formation begins in response to colleterial fluids injected by the females and not by the actions or chemicals of the larvas. The thin walls that surround the larval chamber measure about 1-2 mm thick. The chamber often appears much larger than is required by the small larva within. Prepupas drop via silken strands to the ground, where they build their cocoons. Under favorable field environmental conditions the life cycles of these sawflies can be completed within 45 days. In some areas where the leaves remain on the trees, up to six generations of adults develop within a year. However, the more typical situation is a single generation with pupation in late fall and emergence of adults timed to the appearance of new spring leaves. Caltagirone found an incredible complex of parasites, inquilines, and hyperparasites (which parasitize the parasites) associated with the galls of this sawfly. These associates included six wasps, a moth, and a weevil. Both the moth and weevil feed on the gall tissue, unless they confront another insect on which they will prey. The moth is particularly important in that it alone can account for over 70% of the mortality of the sawfly larvas.

The characteristic tubercles or lenticels on the surface of the galls of *Pontania pacifica* separate them from the galls of other

Pontania sawflies in California. There are at least five other species of *Pontania* galls on willow leaves in this region. These species are far less common and more difficult to identify than *P. pacifica.*

The midge *"Cecidomyia" salicisverruca* induces integral, monothalamous, succulent, cone-shaped, red and yellow leaf galls on several varieties of willows, particularly *S. lasiolepis.* Most of the galls protrude from the undersides of the leaves. The galls are characterized by centrally located depressions at the tips of the galls, creating what appear to be tiny volcanoes. The galls develop singly or in coalescing groups of four to five individuals. each of which measures up to 5 mm high by about 4 mm wide at the base. The dorsal surface of the leaves bears a dimple or slight depression marking each gall mass beneath. The leaves are often bent or pinched where the galls form. Gall formation begins in early spring in the central coast counties. By autumn the galls turn brown and are generally without their occupants. Willows that retain their leaves later into the fall occasionally bear galls

A. Sawfly apple gall, *Pontania pacifica*

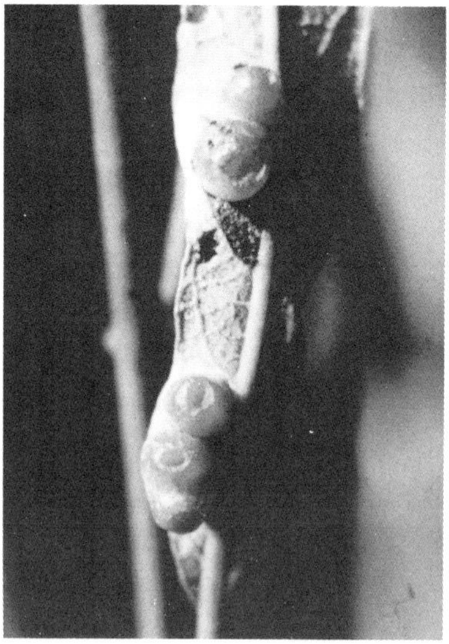

B. Midge gall *"Cecidomyia" salicisverruca*

Fig. 154

that are in good shape at this late date. Although I have never found these galls in large numbers within a small area, they are nevertheless quite common. Little is known about the adults. The quotation marks around the generic name of this midge indicate that this name is the best available and is subject to being changed soon with continuing research.

The willow gall mite, *Aculops tetanothrix,* induces bright red bead galls that protrude on both sides of the leaves of *Salix lasiolepis, S. laevigata,* and others. Each gall normally protrudes on one side of the leaf or the other, but rarely symmetrically on both sides. Galls measure about 1.5 mm high by 1 mm wide, and occur singly or in large clusters numbering several dozen. Heavily galled leaves are often distorted. Galls are commonly covered with a sparse pubescence. Like so many other leaf galls, these make their appearance shortly after new spring leaves develop. Little or no damage results from the actions of this mite and the formation of the galls. The adults or their eggs probably overwinter in the fissures of the bark or around the bud scales.

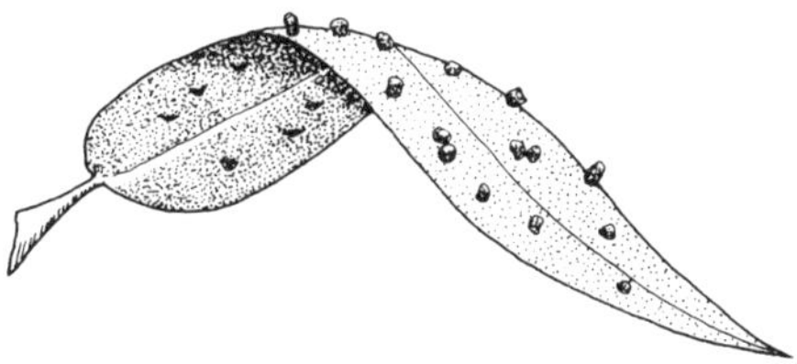

Fig. 155. Mite gall *Aculops tetanothrix*

Shrub Galls

UNLIKE some trees, which may support several species of gall organisms, most shrubs host at most only a few species. Many shrubs are known to be galled by only a single species. In any event, many of the galls found on shrubs are fascinating to study.

Chinquapin Galls

The Sierra or bush chinquapin, *Chrysolepsis sempervirens,* is found among the dry rocky crags of mountainous areas from 600-3350 m from Kern County north through the Sierra Nevada. It is also found in the San Jacinto and San Gabriel Mountains of southern California and along the Coast Ranges from Lake County north into Oregon. This oaklike shrub is easily recognized by the bright rusty-golden hairs on the undersides of the leaves and the spiny, golf-ball-sized fruits. This shrub hosts only one known gall maker, which attacks the flowers.

The chinquapin gall wasp, *Dryocosmus castanopsidis,* induces spherical, monothalamous, golden-yellow to brown, detachable galls on the flowers in late spring and early summer. The round, densely pubescent galls measure 7-13 mm in diameter. The hairs are extremely short and somewhat matted. The hard, solid galls contain a single centrally located larval chamber, which measures about 2 mm in diameter. Frequently one will find only a single gall per catkin, which is usually located near the tip. It is not uncommon to discover several galls growing on the vertical flowers of this shrub. Galls appear to reach maturity by August. Although it is not clear exactly when adults emerge, they most likely make their appearance before or during the time the catkins develop.

The same cynipid wasp also galls giant chinquapin, *C. chrysophylla,* along the coast from Mendocino north to Washington.

Coyote Brush Galls

Any soil left bare in the central and southern Coast Ranges is soon invaded by a tall, hardy, persistent, leather-leaved shrub that dominates the soft, moist chaparral. This is chaparral broom or coyote brush, *Baccharis pilularis,* subspecies *consanguinea.* This common shrub forms dense thickets below 600 m from San Diego to Tillamook, Oregon. The other subspecies of this shrub forms dense mats along the sand dunes and windswept bluffs of the immediate coast. The range of this prostrate form extends from the Russian River in Sonoma County to Big Sur in Monterey County.

In Tilden's exhaustive work on coyote brush he listed 221 species of insects associated with this shrub. However, there are only three common gall organisms that will be discussed here. These are a midge, a moth, and a fungus. All three are distinctive and easily recognized.

The coyote brush gall midge, *Rhopalomyia californica,* induces large, globular, polythalamous, spongy, terminal bud galls on both forms of coyote brush. The soft green to red and

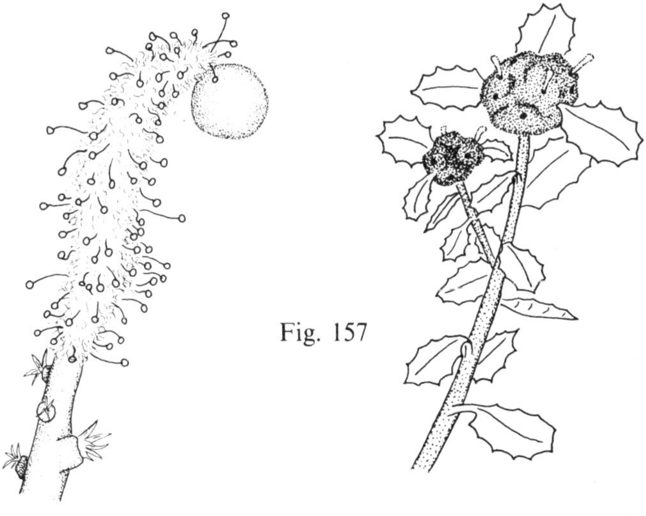

Fig. 157

A. Cynipid wasp gall *Dryocosmus castanopsidis*

B. Midge gall *Rhopalomyia californica.*

purple galls measure up to 2 cm in diameter. The surface of the galls is generally smooth, but characterized by large irregular lumps. Often short vestiges of the leaves protrude from the sides of the galls. The gall mass is usually located at the tips of the branches. Fresh galls can be found at about any time of the year, even though the most growth takes place in the spring months. Each gall contains a number of larval chambers scattered around the center of the gall. Pupal cases are found protruding from the surface of the galls at just about anytime. Adults are known to emerge almost every month of the year in some areas. Tilden found that at least six species of parasites are associated with this midge and its gall. Old galls turn brown and wither.

I have observed specimens of the prostrate subspecies of coyote brush in San Mateo County where not a single terminal bud appeared to have escaped the galling activities of these midges. In such cases the host plant barely resembled coyote brush as it is known. The yellow-green galls form nearly solid carpets over the sand dunes. In the spring new shoots develop from below the dead tissue of the old galls. As these shoots reach upward for light, they too become galled.

The coyote brush gall moth, *Gnorimoschema baccharisella,* induces spindle-shaped, integral, hollow, monothalamous stem galls on both forms of this shrub. The smooth, green, elliptical galls measure up to 3.5 cm long by 1.5 cm in diameter, and occur

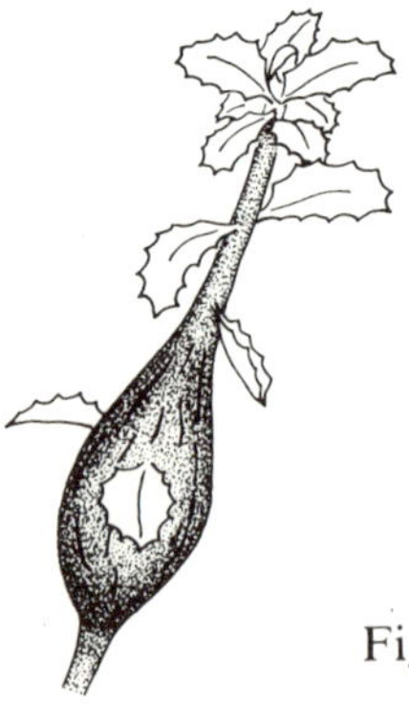

Fig. 158

A. Moth gall, *Gnorimoschema baccharisella.* **B.** *Rhopalomyia californica.* Few terminal shoots of this prostrate form of *Baccharis* remain ungalled in the dunes of the San Mateo coast.

one per shoot. They taper gradually at each end, forming the spindlelike shape. Old galls in September lack the smooth surface of fresh galls and instead reveal shallow fissures. The larval chamber is a massive cavity surrounded by thin walls measuring about 2-3 mm thick. The straw-yellow larva is normally about 5-7 mm long in April in the coastal county areas. By July the larva nearly doubles its length to 10-12 mm. In September the chamber is filled with the frass of the larva. Assuming that the moth larva survives and is not parasitized, it exits through a small hole and drops to the ground via a silken strand. It then pupates among the leaf litter on the ground. The females deposit their eggs in the fall. Gall development does not begin until the following March. A single generation is produced each year.

It is of interest to the ecologist to note that the moist frass serves as a growth medium for various molds, which in turn supply nourishment to fungus-eating insects. As the gall ages, it becomes suitable to a variety of insects. Tilden found at least 10 parasites associated with this gall. He also found that this moth served as a key industry animal producing the most complicated food web of the 221 insects observed on coyote brush. Altogether Tilden found 27 insects interrelating with this moth. The galls of these moths are often found on the same shoots as are the galls of *Rhopalomyia californica.* All of this points to the complexity of interdependence and the diversity of interrelationships that revolve about a single plant species. If we allow our imaginations to multiply this by the number of other plants in a given community and the organisms found therein, the vastness of the ecological spectrum becomes awesome.

The coyote brush rust gall fungus, *Puccinia evadens,* induces integral, elliptical, heavily fissured swellings and massive witches' brooms on both forms of this shrub. The same fungus is also known to gall other species of *Baccharis* in the west. The elongated, nondetachable galls measure up to 18 cm long by 6 cm in diameter. The bright orange uredia develop in the early spring. In some areas these spores can be found at almost any time of the year. Late in the fall the fissures become covered with white,

Fig. 160. Rust gall *Puccinia evadens*

powderlike material where the orange uredia once dominated. The activities of this rust fungus result in large-scale disruption of the energy flow in the host plant. That part of the branch beyond the gall normally dies as a result of this disruption. Galls normally develop on older parts of the shrub.

The witches' brooms that develop usually emanate from the stem swellings. The brooms are composed of dense clusters of short shoots, 15-20 cm long, which die after the first season. Frequently the dense brooms conceal the swollen gall beneath. Old witches' brooms become ideal resting and feeding places for a variety of insects and spiders. The old stem swellings rot easily and crumble with handling.

Pritchard discovered a cecidomyiid midge, *Mycodiplosis pucciniae,* whose larvas feed on only this gall fungus. The larvas actually feed on the contents of the orange spores. These insects significantly reduce the spores available for dissemination.

Creosote Bush Galls

Creosote bush, *Larrea tridentata,* can be seen for miles as it sparsely covers the vast desert flats and slopes. It ranges up to 1500 m throughout southeastern California and into Texas, Utah, and Mexico. This aromatic, tall shrub is characterized by not only the bright yellow flowers and the dark bands around the stems, but also by the golf-ball-sized galls that are easily noticed from the roadside. Actually there are several different galls that appear on this host, but only one is known (see fig. 183A).

The creosote stem gall midge, *Asphondylia auripila,* induces polythalamous, nondetachable, brittle bud galls. New soft, green galls develop in the spring. The ultimate size of the galls is 2.5 cm in diameter. These galls are characterized by long (10 mm), thin (0.5 mm) filaments that form numerous tiny rosette

Fig. 161. Midge gall *Asphondylia auripila*

patterns. The slender filaments rise from the sides of the smaller individual galls that comprise the main gall mass. When stripped of the filaments, the small, pyramidal galls measure 7 mm high by 3 mm wide at the bases. Each of the individual galls is attached at the base to the central hard core of the main gall mass. The funnellike rosette designs are created as the filaments rise and recurve outwardly from the central column. This curious design serves its function well. At the tip of each of the larval chambers a pyramidal plug develops that measures about 2.5 mm long. This plug allows the escape of the prepupas prior to pupation. The plug is forced upward through the neck of the funnel and probably drops to the ground. The prepupas build their cocoons at the entrance to the funnels. These silky-white cocoons measure about 3 mm long and are most noticeable around December. Males and females emerge in late winter and early spring.

Old galls are brown and extremely brittle. At this point they are nearly impossible to remove without destroying the galls, even though they form on the outside of the stems. As these galls develop, they encompass the stems in such a fashion that removal becomes impractical. For this reason, galls remain on the shrubs for several years. Few creosote bushes escape galling by this midge. Other midge galls found on this host are distinguished by the size and pattern of the outer filaments.

Lupine Galls

Of the approximately 82 species of lupines in California, yellow bush lupine, *Lupinus arboreus,* is one of the most elegant. This tall, yellow-flowered shrub is found abundantly near the coast between Ventura and Del Norte counties. Like so many other lupines, clear identification of this shrub requires a technical key.

The lupine bud gall midge, *Dasineura lupini,* induces large, globular, polythalamous, hairy bud galls in the late spring. Those galls found in the shade are usually green, while those growing in the sun are colored red and purple. At maturity the galls measure up to 3 cm long by 2 cm in diameter. The silvery hairs that cover the galls are about 1 mm in length. The gall is composed of overlapping layers of greatly expanded bud scales. Dwarfed, petioled leaflets develop at the tips of the outer scales. In cases where these leaflets do not form, the tips of the bracts recurve away from the gall body. The galls occur singly or in groups of small galls situated close together. The larval chamber is located centrally near the base. Galls collected in July harbor light orange to melon-orange larvas. Adults emerge in May of the following season.

Another, more elliptical but integral stem gall occurs on the same shrub. This is probably the work of the cecidomyiid

Neolasioptera lupini. This gall is evenly tapered at each end and regular in form. A third gall appears irregularly bent and twisted. These integral stem galls are found on the lupines along the Mendocino coast, and are characterized by the general roughness and the bract appendages located along the sides of the rather thin galls. Both forms of these galls measure about 3 cm long by 1 cm in diameter. The manner in which lupine leaves develop in typically compact, dense clusters makes finding these galls difficult at best. Patience and careful observation are key requirements in the discovery of galls and their inhabitants.

A. Midge gall *Dasineura lupini* **B.** Midge gall *Neolasioptera lupini*

Fig. 163

Manzanita Galls

Munz lists over 70 species, subspecies, and varieties of manzanita, *Arctostaphylos* sp., in California. They are found in nearly all plant communities in this area and are frequently the dominant members of chaparral. Manzanita is basically separated from other shrubs by its reddish-brown scaly bark and underlying smooth surfaces, and by its urn-shaped flowers and grey-green erect leaves. The principal gall organism found on manzanitas is an aphid, which is not restricted to a single species of manzanita, but occurs widely in the genus. Aphid galls are common on the species without bristly hairs on the leaves. I have yet to find any galls on those species with bristly pubescent leaves, even though it has been reported that no species of manzanita is immune to this aphid.

The manzanita fold gall aphid, *Tamalia coweni,* induces red, elliptical fold galls along the margins of the leaves. These irritated and swollen sections of the leaves fold over toward the midrib vein, enclosing the stem mothers. The galls normally measure about 10 mm long by 5 mm across. Although they tend to occur singly, it is not uncommon to find both margins of the leaves completely galled with three or more galls on each side. Young leaves are often totally encompassed by gall development in heavily infested areas. Gall formation begins in early spring at low elevations and later at higher altitudes. However, fresh galls can be found at about any time of the year depending on the location. The galls of the Sierra manzanitas are usually dry by autumn, but in the central coastal counties active galls are found well into fall. Frequently the red color of these galls is light enough to reveal the reddish leaf veins within, which appear more like blood vessels. When the galls are gently unfolded, the 2 mm-long stem mothers and offspring can be found amidst a white, cottony wax, which they secrete. The stem mothers are particularly active in the summer and fall, but can be found in an active form throughout the year in some locations.

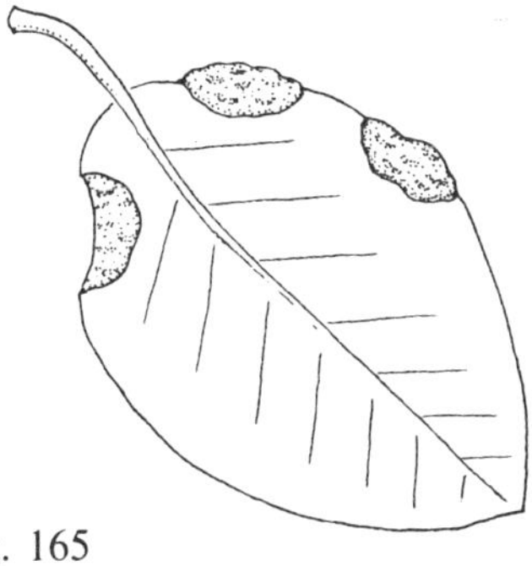

Fig. 165

A. Aphid fold gall, *Tamalia coweni* B. Aphid gall *Tamalia coweni*

Generally damage to the host is minimal. However, some manzanitas suffer tremendous growth setbacks under conditions of heavy infestation where most of the terminal leaves are galled beyond recognition. *Tamalia coweni* is found in the Rocky Mountains as well as in California.

A second aphid gall forms on the flower stalks in place of the flowers. This results in the creation of clusters of folded, round, pea-sized, green flower galls. These galls often eliminate flower and seed production on an individual shrub. The identity of this aphid has not yet been determined due to limited collections. It may prove to be a known aphid, a new species, or the preceding leaf-folding form with diverse galling habits.

Poison Oak Galls

While the delicate, three-leaved pattern and bright fall hues of red, orange, and purple represent the recognizable scourge poison oak, *Rhus diversiloba,* to the susceptible campers and hikers alike, it is of tremendous benefit to wildlife as food, shelter, and habitat. Poison oak is a common shrub in many plant communities below 1500 m throughout California's foothills north into Washington. There is at least one species of eriophyid mite that flourishes on the leaves and possibly flowers of this plant. A third gall forms on the terminal buds of this shrub, which could prove to be a significant biological control mechanism.

The poison oak leaf gall mite, *Aculops toxicophagus,* induces raised, pubescent, bead galls on the surfaces of the leaves in late spring. The tiny, 1 mm in diameter, red beads occur singly or in great coalescing masses resembling **coxcombs** measuring up to 5 mm across. The galls occur on both sides of the leaves. When they form a coxcomb pattern, the galls can distort and bend the leaves. Heavily infested leaves do not attain normal size. The coxcomb galls appear convoluted. Occasionally these mites attack the terminal leaves so severely that they render them unrecognizable.

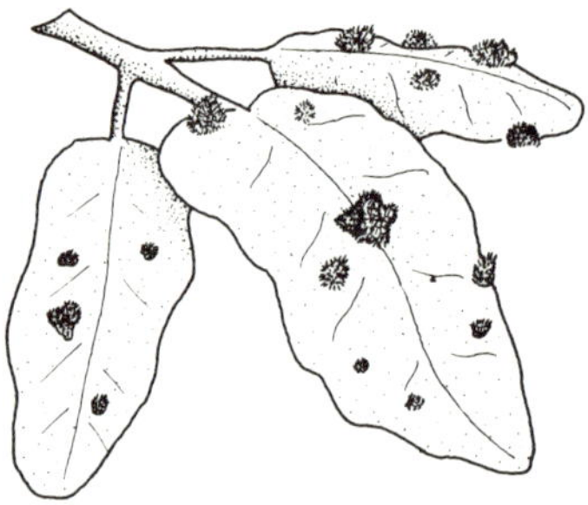

Fig. 166. Mite gall *Aculops toxicophagus*

An eriophyid mite attacks the flowers and prohibits the formation of fruit. The aborted flower tissue fans out in a fasciated manner. Large clumps of these leaflike fasciations develop to about 1 cm across. A clear identification of this mite has not been possible, to date. Possibly a third species of mite attacks the terminal buds of poison oak, causing massive fasciations often 15 cm wide by 25-30 cm long that bend and twist in ways that tease the imagination (see fig. 190B). Distorted, small leaflets develop at the ends of these fasciated galls. The second season of attack reveals new shoots springing from the tips of the fasciation galls. The terminal sections of these shoots also fasciate. After 3-4 seasons of attack by these unidentified organisms, the branches bearing the galls usually die. Entire patches of poison oak have been killed by the persistent attacks of these organisms.

Rose Galls

Several species and varieties of wild rose occur in California and adjacent states. Since there appears to be little host specificity among the gall insects associated with members of the genus *Rosa,* there is no need to distinguish the species of this host plant group. The principal gall insects on roses are cynipid wasps of the genus *Diplolepis.*

The rose leaf gall wasp, *Diplolepis polita,* induces round, spiny, monothalamous, pea-sized leaf galls on *Rosa gymnocarpa* and *R. californica* among others. These detachable, spring galls measure up to 10 mm in diameter exclusive of the spines, which may be as long as the width of the galls. The galls are

Fig. 167. Cynipid wasp gall *Diplolepis polita*

typically red or yellowish-green when shaded. They occur singly or in massive clusters that completely obliterate supporting leaf tissue. Galls that grow close together often coalesce forming potato-shaped, spinose masses of 20 mm or more in length. Often these galls appear as stem galls due to the absence of recognizable leaf tissue. When fresh, the spines are flexible and straight. With age, the spines become curved and brittle, and are easily dislodged with handling. A normal, nonparasitized gall has a fairly large larval chamber with only a thin, smooth wall of tissue forming the gall. Parasitized galls frequently have several larval chambers or a single chamber with an irregular inner wall. The normal galls, as described above, are referred to as the **spring-initiated** galls. These can be compared with galls induced by the same wasps, which develop 60-70 days after oviposition, which is much later than the spring galls. These galls are known as **sucker shoot** galls. They are easily distinguished from the spring galls because they are located along the midrib veins of leaves on new shoots. Sucker shoot galls also bear long, fine, hairlike spines, which are generally oriented in an upward direction.

As with all members of this genus, the larvas are responsible for gall formation. Fecal matter does not build up in the larval chamber because the gut of the larva is closed. As the galls harden and dry in September, the larvas phase into their normal diapause, and they spend the winter in this condition. The galls usually drop to the ground by early autumn. Pupation takes place in the galls, and the adults chew their way out in the early spring. Although males are known for this species, most of the reproduction seems to be parthenogenetic with some sexual mating taking place. At least five species of insects have been found by Shorthouse to be associated with this wasp and its gall.

The rose stem gall wasp, *Diplolepis californica,* induces large, round, polythalamous, leafy stem galls on roses in the spring. These massive green to red galls are characterized by lobed, leaflike projections covering the surface of the galls. These galls

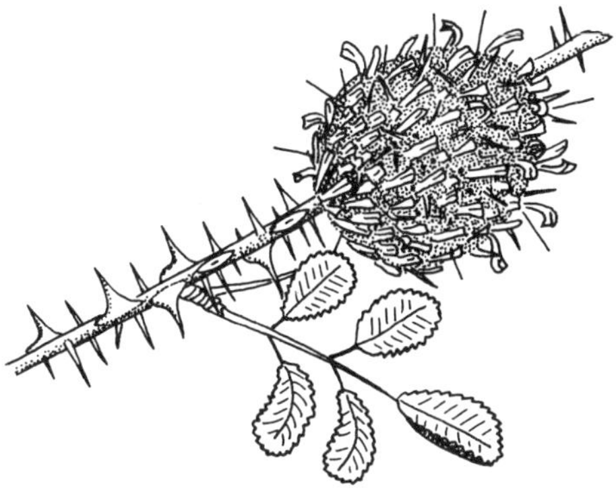

Fig. 169. Cynipid wasp gall *Diplolepis californica*

measure up to 6 cm long by 4 cm in diameter. The usual size is closer to 3 cm with a more rounded shape. When fresh, the red and green lobes are from 2-5 mm long and flexible. As the gall ages, the leaflike lobes shrivel and become brittle. The galls normally completely encompass the stems. The larval chambers measure about 3 mm long. Adults emerge in early spring. Galls are noticeable by May, and reach a peak in their growth by mid to late June in the central coastal counties. This species does not appear to be as abundant as *D. polita.* One of the best locations I have ever found for studying these lobed and spiny galls as well as those of *D. polita* was carelessly sprayed by a county road crew, destroying the once flourishing population. To this date, nearly three years later, there has been little recovery. These wasps pose absolutely no threat to ornamental roses.

Saltbush Galls

There are more than 30 species of saltbush, *Atriplex* sp., in California. I have found galls on only one species so far. Fourwing saltbush, *Atriplex canescens* var. *laciniata,* is a common shrub of the Mohave Desert and the valleys of San

Benito, San Luis Obispo, and San Diego counties. Distinguishing this species from others requires the careful use of a technical key like that of Munz's. A single gall organism is known to occur on this shrub.

The saltbush gall midge, *Asphondylia atriplicis,* induces large, globular, white-woolly, detachable galls on the stems of fourwing saltbush. In the spring the clean, white mass readily stands out amongst the grey-green foliage of the host plant. The galls measure about 2.5 cm in diameter, and usually encompass the entire stem. The white hairs measure 5-7 mm long and are extremely compact on these polythalamous galls. The hairs do not peel off the gall body easily, but instead flake off in short, 1 mm-long bits with persistent rubbing. The larval chambers have thick walls, are elliptical in shape, and measure about 3 mm in length. Little else is known about this creature. Its habits and life cycle need further study.

Fig. 170. Midge gall *Asphondylia atriplicis*

Sage Galls

Of the 19 or so species of sage, *Salvia* sp., recognized in California, only two species seem to host gall organisms. Black sage, *Salvia mellifera,* and white sage, *S. apiana,* support two forms of what appear to be galls induced by the same midge. Black sage is found in the Coast Ranges from Contra Costa and west Stanislaus counties south to northern Lower California. These sages are distinguished from sagebrush, *Artemisia* sp., by the broader, thicker leaves and the insect-pollinated flowers of *Salvia.*

Although the two forms of the galls are considered to be induced by the same midge, further research may indicate distinct species. But for the time being the two will be described as one species.

The sage gall midge, *Rhopalomyia salviae,* induces monothalamous, conical, thick-walled, nondetachable leaf galls on both sages and possibly others. These succulent, green and red galls are covered with matted hairs, and they protrude on both sides of the leaves. The galls occur singly or in close groups along

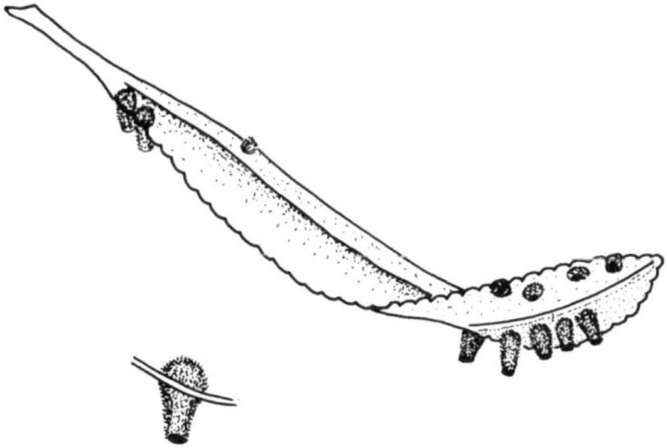

Fig. 171. Midge gall *Rhopalomyia salviae*

the margins or midrib vein and on the petioles. When in large numbers, the galls deform the affected plant parts. The individual galls measure up to 12 mm long by 8 mm wide at the base or point of connection with the leaf surface. On one side of the leaf the galls are round, while on the opposite side the galls narrow into a typical volcanolike column with central depressions at the tips. These fuzzy depressions are actually hairline columns or tubes that extend down into the larval chambers. It is through these tubes that the midge makes good its escape. Dull red-brown larvas, 1 mm long, are found in the galls in March around Morro Bay. Specimens collected in San Diego County produced adults in April.

The second form of gall differs mainly in size. These galls measure up to 6 mm long by 1.5 mm in diameter. Although they occur singly, they are more commonly found growing side by side along both edges of the leaves. They often fuse together, creating continuous masses of gall tissue that distort the leaves drastically. Like the larger form, the smaller galls are pubescent, particularly around the apical vents. These galls are thin-walled, more tubular in design, and yellow-green in color. Adults have emerged from galls in rearing containers in May.

Sagebrush Galls

There are two members of the genus *Artemisia* that host galls of interest. One is the California sagebrush, *A. californica,* and the other is the basin sagebrush, *A. tridentata.* California sagebrush is found below 760 m from Marin and Napa counties to southern California in coastal sage scrub and coastal strand communities. The leaves of this sagebrush are divided into linear, threadlike segments. *A. tridentata* grows between 450-3200 m in dry places along the western edge of the deserts and through the Sierra to Modoc and Siskiyou counties and into British Columbia. The leaves of this shrub are flat, pubescent, and divided at the tips into three distinct lobes. Felt recorded several species of gall midges associated with this shrub, but I have found only three in California.

The California sagebrush gall midge, *Rhopalomyia floccosa,* induces white-woolly, monothalamous, thin-walled bud galls on *A. californica.* These late spring and early summer galls develop singly or in dense clusters, which create the appearance of one large gall. Individual galls measure up to 5 mm in diameter, and are covered with long, white hairs. The naked gall is quite small, measuring only about 1-2 mm in diameter. The galls are difficult to detach from the stems. Adults have been reared in November. Extensive galling does result in damage to the overall health of the individual shrub, as observed in Morro Bay, where one shrub was found supporting several hundred galls. This situation, however, does not appear to be common according to my field observations.

Fig. 173. Midge gall *Rhopalomyia floccosa*

The midge *Rhopalomyia* sp. induces large, globular, white-woolly, polythalamous, detachable bud galls on *A. tridentata.* These spring galls measure 2-3 cm in diameter, and consist of hundreds of threadlike, elongate, densely pubescent, white to green lobes. The entire mass has a definite spongy character. The individual lobes are 5-8 mm long and are light green with reddish margins. The well-hidden larval chambers are small and difficult to locate. The galls often completely surround the stems upon which they develop. This species is quite common in San Diego

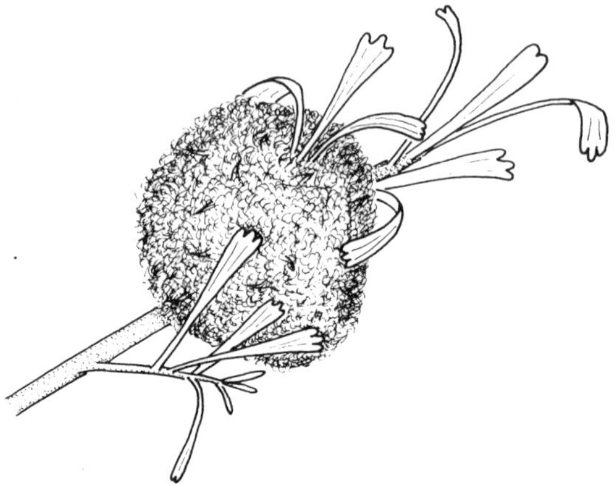

Fig. 174. Midge gall *Rhopalomyia* sp.

County. Felt used the name *Rhopalomyia navasi* in his key. But, since this is a European species, it is doubtful that the same name applies to the North American galls. As with so many other cecidomyiids, there is much need for intensified study in order to clarify the taxonomy of these midges.

Silk Tassel Bush Galls

Silk tassel, *Garrya fremontii,* inhabits the dry, brushy slopes up to 2300 m in the Sierra and along the Coast Ranges south to Monterey County. Its characteristic long tassellike clusters of flowers and thick, glossy leaves set this shrub apart from other hillside shrubs. A single gall organism is known to occur on this shrub.

The cecidomyiid midge *Asphondylia garryae* induces soft, fleshy, monothalamous, detachable, green bud galls on silk tassel. When fresh in June and July in the Sierra, the enlarged bud scales overlap in a most fascinating pattern. The tips of the outermost scales flare away from the gall body. Near the apex of the galls these fleshy bracts combine to resemble lips. The bracts are difficult to separate as they are usually connected about their midlevel to the others. The fresh green, succulent gall measures about 14 mm in diameter. With age, the galls shrivel and turn

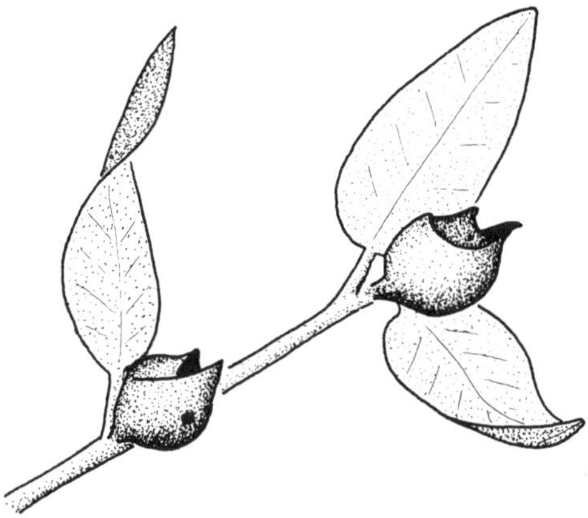

Fig. 175. Midge gall *Asphondylia garryae*

brown or black. The once-obvious bud scales all but disappear with age. The galls bear a sparse coating of fine hairs over the surface. The galls occur singly near the tips of branches. The larval chamber is centrally located in the solid, spongy mass, and measures about 2.5 mm in diameter. Adults have emerged from galls collected in Lake County in May. Adults from Sierra galls emerge later.

Spiraea Galls

Meadowsweet or *Spiraea,* particularly *S. densiflora,* is a common shrub in the moist rocky places in conifer forests from Tulare County north through the Sierra and into the Siskiyou Mountains and Trinity County. This shrub is found up to an elevation of 3350 m, and is recognized by its soft, green-grey, toothed leaves and clusters of tiny pink flowers. I have found only one gall on this host, and it is induced by a midge.

The spiraea pod gall midge, *Rhabdophaga salicifolia,* induces monothalamous, thickened, leaf fold galls on new spring leaves. The rosy-green, succulent swellings measure up to 2.5 cm long by 3 mm in diameter. The cigar-shaped galls form either along the midrib vein, bringing the edges of the leaf together, or at the edges of the leaves with the roll moving toward the midrib vein. It

Fig. 176. Midge gall *Rhabdophaga salicifolia*

is not uncommon to find both edges of a leaf rolled toward the midrib vein. This midge induces both fold and roll galls depending on which part of the leaf is affected. The central fold along the midrib vein is quite common. This fold is made so that the dorsal surfaces of the affected leaf come together. Like other midge fold or roll galls, the spiraea pod gall can be carefully opened along the same axis as the midrib vein. With age, the galls become somewhat brittle. Each gall comprises a single chamber within which are located several bright orange larvas. Galls are fully developed and with mature larvas by August. The activities of the adults are not known. One can speculate that emergence takes place in late spring as new growth begins to develop at the tips of the branches. Also, the adult cycle is probably short-lived, producing a single generation per year. However, much needs to be learned about host-parasite relationships. This midge is quite common throughout the range of the host plant.

Thimbleberry Galls

Western thimbleberry, *Rubus parviflorus,* is a common shrub reaching a height of 1.8 m along the coastal counties, while reaching only about 0.6 m in the Sierra Nevada. Its broad, soft, palmate leaves combine with the large, petaled, white flowers and bright red berries, shaped like thimbles, to create distinctive and unmistakable characters for easy identification. The shrub extends from San Diego County north through the Sierra and Coastal Ranges into Oregon and Washington. While Felt listed several species of gall organisms occurring on members of this genus, western thimbleberry has been found by this writer to be the only representative that hosts a gall organism with any regularity.

The thimbleberry stem gall wasp, *Diastrophus kincaidii,* induces integral, polythalamous, knobby stem galls on new spring growth. The irregularly shaped galls measure up to 8 cm long by 2.5 cm in diameter. In some cases the galls are almost round, while in others they are nearly potato-shaped. The general shape is influenced by the number and location of the growing larvas. When fresh in late spring and early summer, the galls are soft and green. Fresh galls have been found at Point Reyes National Seashore in October and at 1500 m in the Sierra in August. The galls are covered with short, erect hairs. As the galls turn brown in the fall, these hairs become brittle and are easily detached. The surface of the galls often consists of noticeable lumps or knobs. The growth of the galls will occasionally cause the stems to bend radically. The larval

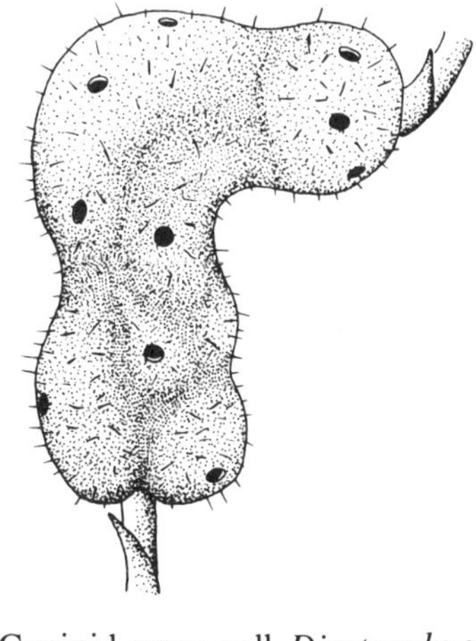

Fig. 177. Cynipid wasp gall *Diastrophus kincaidii*

chambers are situated radially around a thick central stem core and are quite close to the surface. The first signs of gall development are evident about 10 days after egg deposition. The dozens of larvas within a single gall pass through five instar stages during the summer. By fall they enter their long diapause. Adults emerge the following spring. Studies indicate that males emerge prior to females. Wanberg has found that little, if any, flying takes place as adults wander about the foliage seeking new spring growth. After mating, the females deposit their eggs, making several punctures close together. Wangberg also found that a single gall can result from the actions of the larvas hatched from the eggs of two or three separate females that selected the same general area for oviposition. Populations of this wasp are extremely localized, abundant in some areas and totally absent in other areas where the host plant flourishes.

Wild Cherry and Plum Galls

Three members of the **stone fruit** group of native shrubs in California support interesting galls. Bitter cherry, *Prunus emarginata,* western choke cherry, *P. virginiana* variety *demissa,* and Sierra plum, *P. subcordata,* commonly host gall organisms. Bitter cherry occurs below 2700 m from San Diego County north into British Columbia in conifer areas. Sierra plum is found in the Coast Ranges from Santa Cruz County north to Del Norte and Siskiyou counties. In the Sierra this shrub ranges from Tulare and Kern counties below 1800 m north through Modoc County and into Oregon. Western choke cherry is found in the mountains from San Diego County north through the Coast Ranges and the Sierra Nevada into Washington and Idaho in damp places below 2500 m. The subtleties that separate these species of *Prunus* are best found in Munz's work. The galls found on these shrubs involve three different species of fungi and an eriophyid mite. I have found a fourth organism, a midge, to commonly gall the fruit of *P. virginiana* variety *melanocarpa* along the rivers of Montana and Wyoming. But, it appears to be rare here (see Appendix B).

The bitter cherry gall mite, *Phytoptus emarginatae,* induces clublike, beige pouch galls on the leaves of all three shrubs. These blunt-tipped, minutely pubescent galls measure up to 5 mm in height by about 1 mm in diameter. The galls occur singly on both surfaces of the leaves, and several dozen galls per leaf is generally the rule rather than the exception. At the base of the clublike gall on the opposite side of the leaf there is a microscopic opening leading into the gall chamber. When fresh, the galls are beige, green, or red depending on their exposure. Although the galls tend to stand erect, some bend and twist. There is a single generation of these mites per year. Females retire to crevices in the terminal branches in late summer. They carry sperm with them through the winter after having contacted males. The females induce galls the following spring that contain both males and females. This mite is widespread across North America and in Europe. Damage to the leaves and the shrub appears to be minimal. Some browning of the leaf tissue does occur under conditions of heavy infestation.

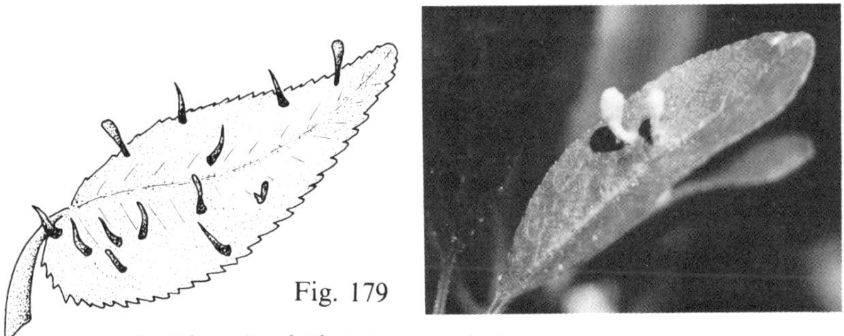

Fig. 179

A. Mite galls of *Phytoptus emarginatae* (pointed nail forms)
B. Club-shaped form of *P. emarginatae*

The bladder plum gall fungus, *Taphrina pruni-subcordatae,* attacks the fruit of Sierra plum. The infected fruit expands and elongates considerably over its normal size of 1.5 cm in diameter. The bladder plum gall develops to about 4.5 cm long by 2 cm. in diameter, and is round to potato-shaped. When fresh in the spring, these green, succulent, wrinkled galls look like large fruit

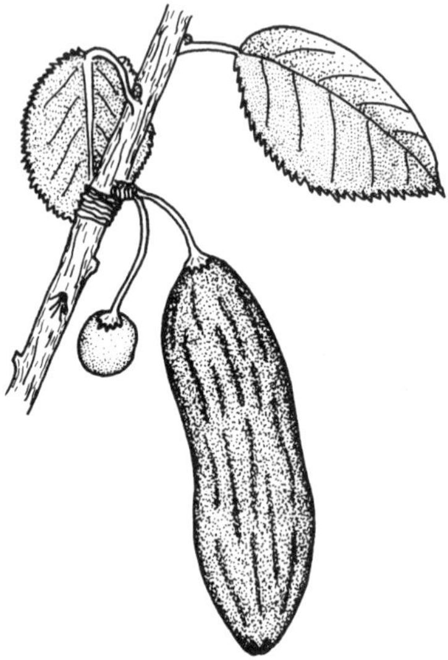

Fig. 180. Fungus bladder plum, *Taphrina pruni-subcordatae*

with their own stems. With age, these galls, commonly referred to as **bladder plums** or **plum pockets,** turn beige, dry, and become hollow. The surface of the galls at this stage becomes somewhat mealy with microscopic pits and bumps and a waxy bloom. Although one might expect thorough and heavy infestations of this fungus to eradicate the production of fruit and seedlings, this does not appear to be the case. While this fungus has a broad range in the west, it tends to be locally distributed. It is not difficult to spend hours looking for these galls without any luck. Then, all of a sudden, there they are as bold as anything can be.

The cherry leaf curl fungus, *Taphrina flectans,* induces massive, dense clusters of swollen, distorted, and discolored leaf curl and witches' brooms on bitter cherry. These yellowish-green galls stand out readily as seen from roadsides. Their compact, discolored nature contrasts greatly with uninfected branches. The witches' brooms bear leaves that are pale green with tints of yellow. The mass of shoots forming the brooms supports hundreds of the altered leaves. Infected leaves have a powdery-waxen bloom on the surface. Severe infestations of this fungus probably seriously affect flowering and the production of fruit, as with peach leaf curl, *Taphrina deformans*. Cherry leaf curl occurs on these native shrubs throughout California, Oregon, Washington, and Idaho. The bitter cherry shrubs along Highway 108 near the Clark Fork of the Stanislaus River in the Sierra support thousands of these brooms with leaf curl.

The black knot fungus, *Dibotryon morbosum,* induces large, black, knobby, pitted galls on several species of *Prunus* including western choke cherry. These elongated, massive swellings of the stems are found at any time of the year as old galls remain on the shrubs for several seasons. Development of the galls begins before new spring leaves appear. In mid-March light, bark-colored swellings develop on the sides of the stems. In the early stages of development the infected stems appear to erupt or explode under the pressure of the growing galls. The growth of the galls on the sides of the stems often causes the stems to bend

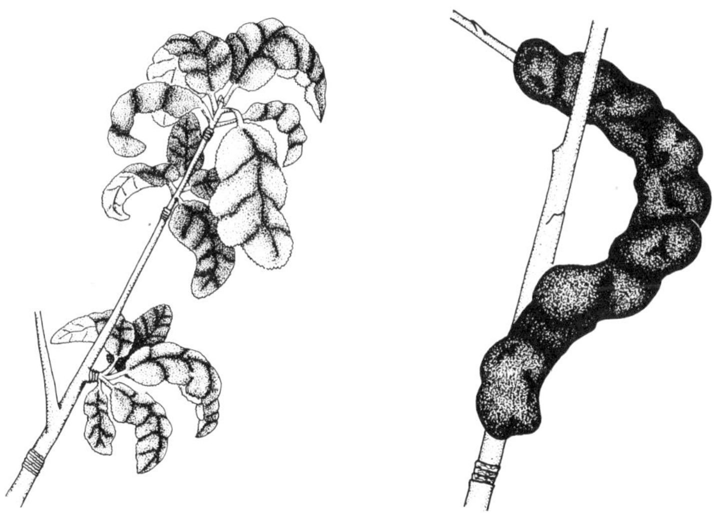

A. Cherry leaf curl, *Taphrina flectans* **B**. Fungus stem gall, *Dibotryon morbosum*

Fig. 181

radically. By summer the galls assume a pale green color. Next, an olive-green, velvety, smooth mat develops over the surface. As the galls mature, they turn black, signaling the completion of spore production. Old galls are rough, pitted, cracked, knobby masses that make excellent homes for small beetles, springtails, and other insects. These galls commonly reach 9 cm in length by 3 cm in diameter. The development of these galls generally kills those parts of the branches above the galls. New galls form just below old galls or on previously uninfected branches. These nondetachable stem galls remain on the branches for several years. In Berkeley I have found a solid patch of western choke cherry, 7.6 m long by 3 m wide, with not a single plant free of this fungus. Countless thousands of black knot galls were found in this patch, which continues to thrive despite the attack of the fungus.

Appendix A: HOST-GALL INDEX

HOST	GALL	ORGANISM
TREES		
Alder	cone bracts	fungus, *Taphrina occidentalis,* p. 78
	leaf beads	mite, *Phytoptus laevis,* p. 79
Aspen	bud	mite, *Eriophyes parapopuli,* p. 86
	leaf erineum	mite, *Phyllocoptes didelphis,* p. 88
Cottonwood	leaf	aphid, *Pemphigus populi-venae,* p. 85
	petiole	*P. populicaulis,* p. 84
		P. populitransversus, p. 82
		P. bursarius, p. 86
		P. nortoni, p. 86
Cypress	bud	midge, *Walshomyia cupressi,* p. 80
Juniper	bud	midge, *Oligotrophus betheli,* p. 89
	bud	midge, *Walshomyia sabinae,* p. 89
Maple	leaf erineum	mite, *Eriophyes calaceris,* p. 92
Oak, black	flower	wasp, *Callirhytis congregata,* p. 95
	flower	*C. serricornis,* p. 96
	flower/leaf[1]	*Dryocosmus dubiosus,* p. 96
	stem/leaf	*Callirhytis quercussuttoni,* p. 98
	stem	*C. perdens,* p. 100
	stem/leaf[1]	*C. pomiformis,* p. 100
	bud/leaf	*C. agrifoliae,* p. 102
	leaf	*Dryocosmus minusculus,* p. 103
	petiole/leaf/acorn	*Callirhytis flora,* p. 103
	leaf	*Callirhytis perfoveata,* p. 104
	leaf erineum	mite, *Eriophyes mackiei,* p. 105
Oak, white	stem/bud	wasp, *Andricus chrysolepidicola,* p. 106
	stem	*A. californicus,* p. 109
	stem	*Disholcaspis canescens,* p. 108
	stem	*D. corallina,* p. 109
	stem	*D. washingtonensis,* p. 110
	stem	*Besbicus conspicuus,* p. 110
	stem	*Disholcaspis eldoradensis,* p. 110
	stem	*D. plumbella,* p. 112
	stem	*D. prehensa,* p. 112
	leaf	*Andricus crystallinus,* p. 114
	leaf	*A. brunneus,* p. 115
	leaf	*A. kingi,* p. 116
	leaf/flower[1]	*A. gigas,* p. 117
	leaf	*A. fullawayi,* p. 117
	leaf	*A. stellaris,* p. 118

1. When two plant parts are listed, this indicates that the species has an alternate generation with separate galls occurring on these organs.

HOST	GALL	ORGANISM
	leaf	*A. pattersonae*, p. 119
	leaf	*A. confertus*, p. 120
	leaf	*Sphaeroteras trimaculosum*, p. 120
	leaf	*Neuroterus saltatorius*, p. 120
	leaf/bud[1]	*Antron echinus*, p. 122
	leaf/bud[1]	*A. douglasii*, p. 123
	leaf	*A. clavula*, p. 124
	leaf/bud[1]	*Andricus atrimentus*, p. 125
	leaf	*Dros pedicellatum*, p. 126
	leaf	*Andricus parmula, p. 118*
Oak, canyon live	stem	wasp, *Dryocosmus asymmetricus*, p. 127
	stem/leaf	*Heteroecus dasydactyli*, p. 128
		H. melanoderma, p. 128
	stem/leaf[1]	*H. pacificus*, p. 130
	stem	*H. sanctae-clarae*, p. 132
	stem	*Disholcaspis chrysolepidis*, p. 132
	stem	*Andricus spectabilis*, p. 134
	stem	*Disholcaspis truckeensis*, p. 134
	stem	*Andricus reniformis*, p. 135
	stem	*Trichoteras vaccinifoliae*, p. 136
	leaf	*Andricus lasius*, p. 137
	leaf	*Paracraspis guadaloupensis*, p. 138
	leaf	*P. patelloides*, p. 138
	leaf	mite, *Eriophyes mackiei*, p. 139
Pine	stem	fungus, *Peridermium harknessii*, p. 140
	stem	plant, *Arceuthobium campylopodum*, p. 141
	stem	*A. americanum*, p. 141
Poplar	petiole	aphids, *Pemphigus spp.*, pp. 82-86
Spruce	stem	adelgid, *Adelges cooleyi*, p. 142
Walnut	leaf erineum	mite, *Eriophyes erineus*, p. 144
	leaf	mite, *E. brachytarsus*, p. 144
Willow	stem	wasp, *Euura exiguae*, p. 146
	stem	*E. geyerianae*, p. 147
	stem	*E. scoulerianae*, p. 148
	stem	*E. breweriae*, p. 148
	stem	*E. lemmoniae*, p. 149
	stem	*E. lasiolepis*, p. 150
	bud rosette	midge, *Rhabdophaga brassicoides*, p. 150
	stem	midge, *R. salicisbatatus*, p. 152
	leaf	wasp, *Pontania pacifica*, p. 153
	leaf	midge, *"Cecidomyia" salicisverruca*, p. 154
	leaf	mite, *Aculops tetanothrix*, p. 155

1. When two plant parts are listed, this indicates that the species has an alternate generation with separate galls occurring on these organs.

HOST	GALL	ORGANISM
SHRUBS		
Chinquapin	flower	wasp, *Dryocosmus castanopsidis,* p. 156
Coyote brush	bud	midge, *Rhopalomyia californica,* p. 157
	stem	moth, *Gnorimoschema baccharisella,* p. 158
	stem	fungus, *Puccinia evadens,* p. 159
Creosote brush	bud	midge, *Asphondylia auripila,* p. 161
Lupine	bud	midge, *Dasineura lupini,* p. 162
	stem	midge, *Neolasioptera lupini,* p. 163
Manzanita	leaf	aphid, *Tamalia coweni,* p. 164
Plum, Sierra	fruit/leaf	fungus, *Taphrina* sp., p. 179
Poison oak	leaf	mite, *Aculops toxicophagus,* p. 166
Rose	leaf	wasp, *Diplolepis polita,* p. 167
	stem	*D. californica,* p. 168
Sage	leaf	midge, *Rhopalomyia salviae,* p. 171
Sagebrush	bud	midge, *R. floccosa,* p. 173
	bud	midge, *Rhopalomyia* sp., p. 173
Saltbush	bud	midge, *Asphondylia atriplicis,* p. 170
Silk Tassel	bud	midge, *Asphondylia garryae,* p. 174
Spiraea, Meadowsweet	leaf	midge, *Rhabdophaga salicifolia,* p. 175
Thimbleberry	stem	wasp, *Diastrophus kincaidii,* p. 177
Wild cherry	leaf pouch	mite, *Phytoptus emarginatae,* p. 179
choke	leaf curl	fungus, *Taphrina flectans,* p. 180
bitter	fruit	fungus, *T. pruni-subcordatae,* p. 179
	stem	fungus, *Dibotryon morbosum,* p. 181

Appendix B: MISCELLANEOUS GALLS

The following is a composite of gall organisms in their respective host plants that are known to occur in California. These were not included in the general descriptions because of either their rarity, absence in the author's collections, lack of adequate descriptive material, or because the hosts are neither shrub nor tree.

HOST	GALL	ORGANISM
Woodfern, *Dryopterus arguta*	gelatinous gall on leaflets	fungus, *Taphrina californica*
Fremont cottonwood, *Populus fremontii*	concave/convex swollen leaf gall	fungus, *Taphrina populi-salicis*
Black cottonwood, *Populus trichocarpa*	fasciated catkin	mite, *Eriophyes neoessigi*
Holly leaf cherry, *Prunus ilicifolia*	witches' broom	fungus, *Taphrina thomasii*
Desert peach, *Prunus andersonii*	erineum gall	mite, *Eriophyes prunandersoni*
Meadowsweet, *Spiraea densiflora*	flower gall	mite, *Eriophyes spiraea*
Buckbrush, *Ceanothus velutinus*	leaf bead gall	mite, *Eriophyes ceanothi*
Ceanothus sp.	bud gall	midge, *Asphondylia ceanothi*
Western ragweed, *Ambrosia psilostachya*	leaf bead gall	mite, *Eriophyes boycei*
Sagebrush, *Artemisia californica*	leaf erineum	mite, *Eriophyes paracalifornica*
	conical leaf gall	midge, Rhopalomyia *californica*
Seep willow, *Baccharis glutinosa*	leaf bead gall	mite, *Eriophyes baccharices*
Baccharis pilularis	integral stem gall	midge, *Rhopalomyia baccharis*
Sea blite, *Suaeda californica*	globular leaf gall	midge, *Asphondylia dondiae*
California fuchsia, *Zauschneria californica*	rosette gall	midge, *Contarinia zauschneriae*

HOST	GALL	ORGANISM
Fiddleneck, *Amsinckia lycopsoides*	flower gall	midge, *Schizomyia macrofila*
Vervain, *Verbena lasiostachys*	floral stem gall	midge, *Neolasioptera verbenae*
Penstemon antirrhinoides	stem gall	midge, *Lasioptera tibialis*
Monkey flower, *Mimulus longiflorus*	woolly bud gall	midge, *Asphondylia diplaci*
Mimulus aurantiacus	stem gall	midge, *Neolasioptera mimuli*
Honeysuckle, *Lonicera subspicata*	bud gall	midge, *Rhopalomyia lonicerae*
Snowberry, *Symphoricarpos mollis*	midrib leaf fold gall	midge, *Resseliella californica*
Gumweed, *Grindelia humilis*	flower gall	midge, *Rhopalomyia grindeliae*
Grindelia robusta	blister leaf gall	midge, *Asteromyia grindeliae*
Haploppapus palmeri	rosette gall	midge, *Rhopalomyia ericameriae*
Rabbit brush, *Chrysothamnus nauseosus*	bud gall	midge, *Rhopalomyia bigelovioides*
Tetradymia comosa	woolly bud gall	midge, *Mayetiola tetradymia*
Western chokecherry, *Prunus virginiana*	swollen fruit gall	midge, *Contarinia virginianae*
Spruce, *Picea* sp.	swollen, clustered needle gall	adelgid, *Pineus pinifoliae*
Blue oak, *Quercus douglasii*	puffy, globular erineum	mite, *Eriophyes trichophila*

The following list of galls and gall organisms is largely taken from Weld's 1957 work. It is presented here only to more accurately reflect known cynipid wasp populations as discussed by Weld. I have not explored any synonomy with these names; thus taxonomic changes may be in order. The galls are presented according to the three generic groups of oaks rather than to the species.

HOST	GALL	ORGANISM
White oaks	flower galls	wasp, *Neuroterus floricola*
		wasp, *Neuroterus florulentus*
		wasp, *Dryocosmus gigas*
	acorn galls	wasp, *Neuroterus cupulae*
		wasp, *Andricus albobalani*
	integral stem gall	wasp, *Neuroterus varians*
		wasp, *Andricus wiltzae*
	detachable stem gall	wasp, *Disholcaspis simulata*
		wasp, *Disholcaspis conalis*
		wasp, *Disholcaspis mamillana*
		wasp, *Disholcaspis mellifica*
	bud gall	wasp, *Andricus occulatus*
		wasp, *Andricus opertus*
		wasp, *Andricus verensis*
		wasp, *Andricus chico*
	detachable leaf gall	wasp, *Besbicus mirabilis*
		wasp, *Besbicus maculosus*
		wasp, *Xanthoteras teres*
		wasp, *Andricus albicomus*
		wasp, *Trichoteras tubifaciens*

HOST	GALL	ORGANISM
		wasp, *Andricus discularis*
	integral leaf gall	wasp, *Andricus fimbrialis*
		wasp, *Neuroterus decipiens*
		wasp, *Neuroterus fragilis*
		wasp, *Dros atrimentum*
		wasp, *Neuroterus engelmanni*
		wasp, *Neuroterus washingtonensis*
		wasp, *Andricus stellulus*
Black oaks	flower gall	wasp, *Callirhytis floripara*
	acorn gall	wasp, *Callirhytis milleri*
		wasp, *Eumayria eldoradensis*
		wasp, *Callirhytis carmelensis*
	bud gall	wasp, *Callirhytis uvellae*
	detachable leaf gall	wasp, *Dryocosmus grumatus*
Canyon live oak/ Huckleberry oak	acorn gall	wasp, *Loxaulus trizonalis*
		wasp, *Andricus chrysobalani*
	bud gall	wasp, *Andricus projectus*
		wasp, *Trichoteras frondeum*
	detachable leaf gall	wasp, *Trichoteras coquilletti*
		wasp, *Trichoteras perfulvum*
		wasp, *Trichoteras rotundula*

HOST	GALL	ORGANISM
		wasp, *Paracraspis insolens*
		wasp, *Heteroecus flavens*
		midge, *Youngomyia quercina*

Fig. 190**A**. Unidentified species of midge gall on junipers.

Fig. 190**B**. Unidentified species of gall on poison oak.

Appendix C: GLOSSARY

Adelgids. Insects closely related to aphids but in the family Phylloxeridae. Although they are commonly referred to as aphids, a misnomer, they are correctly known as adelgids.

Agamic. Referring to the asexual or unisexual female generation, having no sexual union.

Alates. Refers to winged aphids.

Bead Gall. A globular pouch gall with an opening at the bottom.

Bristles. Stiff hairs.

Catkin. A long floral spike of clustered flowers, standing erect or drooping.

Convoluted. Irregular folds and ridges, brainlike.

Crenate. Margin with rounded teeth.

Deciduous. Refers to plants that lose their leaves regularly each year, generally in the late summer and fall.

Diapause. A period of rest in the life of an insect when growth is arrested and metabolism is low.

Dorsal. Relating to the upper surface.

Erineum. Hairy, velvety, or pilelike growths on the surface of leaves, caused by eriophyid mites.

Fasciation. An abnormal enlargement or flattening of plant stems or flower stalks.

Fold Gall. A gall having leaf tissue folded over the insect in a single layer

Frass. Wood or other plant tissue fragments mixed with excrement produced by an insect.

Gall. An abnormal plant growth, swelling, or tumor induced by another organism; includes fruit or buds aborted or altered by gall organisms.

Glabrous. Without hairs

Hyperparasite. A parasite that attacks other parasites.

Inquiline. An insect that feeds on the plant tissue of a gall and not necessarily the gall organism

Instar. The stage of an insect between larval moults.

Integral. A nondetachable bulge or swelling of leaf or stem.

Larva. An immature stage between egg and pupa of an insect with complete metamorphosis; a maggot or caterpillar.

Leaf Curl. A leaf gall in which tissues are swollen, distorted, and discolored; on peaches and plums.

Mark Gall. A detachable or integral leaf gall

Midrib. The central vein of a leaf.

Monothalamous. Galls inhabited by one or more larvas in a common cell, cavity, or chamber.

Oviposition. The act of a female depositing eggs.

Ovipositor. The external, egg-laying apparatus of an insect.

Parasite. An animal or plant that lives in or on another living organism for at least part of its life cycle, at the physical expense of the host.

Parthenogenesis. Reproduction by the female without fertilization by a male.

Petiole. The stem preceding a leaf or flower, but attached to a main branch or stem.

Polythalamous. Galls containing two or more larvas with each in separate cells or chambers

Pouch Gall. A club-shaped or nodular gall characterized by a microscopic opening at the base, caused by mites.

Pubescent. Covered with short hairs.

Pupa. The resting stage of an insect between larva and adult where physical transformation takes place, with those insects with complete metamorphosis.

Roll Gall. Swollen leaf tissue that rolls over the inciting organism with more than one layer; not to be confused with leaf rolling insects that move the leaf tissue mechanically and fix it in place; roll galls are physiologically manipulated

Spermatophores. Slender stalks bearing sperm sacs at the apex. In eriophyid mites the females, after contact with males, carry spermatophores with them through winter

Stem Mother. In aphids the female that initiates gall formation and the production of several generations of offspring.

Stylet. The needlelike part of the piercing-sucking mouthparts of an insect or mite

Succulent. Juicy, fleshy, soft, and spongy.

Tubercle. A small prominence, scale, or nodule.

Umbo. A central lump or conelike knob.

Unisexual. Referring to the female generation when no males are produced.

Uredospores. Summer spores formed in the uredinium of certain rust fungi.

Ventral. Relating to the underside or lower surface

Witches' Broom. An abnormal cluster of shoots or twigs emanating from a common focal point on stems and branches, caused by mechanical injury (not a gall) or by an organism (a true gall).

Appendix D: HOW TO REAR GALL INSECTS

Some groups and species of gall insects are easier to successfully rear out of their gall homes than others. Wasps, for example, will often emerge from their galls more frequently when collected at the right time of year than will midges. If collected towards the end of larval feeding stage or when the insect is in diapause or pupal form, the chances of normal development and adult emergence are much better. The key factor is timing. You can get an idea of what stage the insects are in by cutting open a sample gall of the selected species to determine whether the insect is active or at rest. The second alternative is to wait until the appropriate season for collecting and then take your chance.

When collecting galls for the purposes of rearing the insects, study, and identification, there are some things to consider with each kind.

Wasps. Complex life cycles with alternate generations, and unknown information on biology and the effect of environmental changes make many wasps difficult, if not impossible, to rear. However, many others rear out of their galls quite readily. Success can be achieved with many of the spring leaf and flower galls on coast live oaks by collecting in April or May, depending on the location. Other summer wasp galls issue their inhabitants during the following spring. To be safe, these galls should be collected in the late fall months. Blue oaks are excellent sources of a variety of summer galls. Some high-altitude oak galls can be collected in the fall and, if placed in a refrigerator from late October through March, will often issue their adult inhabitants. Some galls, like the California oak apple, will slowly release a variety of insects, including parasites, for two or three years after collection. The diversity is quite exciting.

Aphids. These insects can be easily found by dissecting the galls and removing them for microscopic study. Summer and early fall months are good times to hunt gall aphids.

Midges. These insects are the most difficult group to rear. I have had luck with only a few species. But with patience and perseverance on the part of many people the problems will be worked out and information will accumulate. The midges that gall creosote brush and coyote bush can be reared, if you collect the galls while the insects are in their cocoons at the surface of the galls.

Collection. Freshly collected galls should be placed in glass or plastic jars with screened tops for easy viewing and ventilation. As many of the leaves as possible should be removed to prevent excess humidity. Both succulent, juicy leaves and air-tight containers are conducive to mold, which obliterates the galls and destroys the insects.

Monitoring. The short lifespan of many gall insects requires frequent checking in order that you find the adults alive. It is more important to observe, study, experiment, and release live gall insects than to find the brittle, crumbled bodies on the bottom of the rearing container. For wasps, you might use fine screening or cheesecloth to cover a containerized oak seedling of the right species to try to transplant galls through the actions of the insects you have reared. The months of patience required has proved worthwhile for many scientists who have perfected this technique.

Discretion. Although it has been suggested that it is difficult to endanger an insect species through over-collection, it is clearly inappropriate to collect 109,218 *Antron echinus* galls to rear the adults. No one need strip a tree or area of a particular gall species in order to study it. In fact, too often some animals are studied to the brink of extinction. Discretion in collecting is a sound conservation measure to preserve a species in its natural habitat while gathering enough to learn its habits.

I would be remiss to suggest that it is easy to rear gall insects. Sometimes, no matter how careful you are in timing the collection or caring for the galls, the insects within perish before maturing. Trial and error, taking careful notes, and experimenting with different techniques can help to reveal some of the secret lives of gall insects.

Bibliography

Arthur, Joseph C. 1962. *Manual of the rusts in U.S. and Canada.* New York: Hafner.

Ballou, Lois M. 1925. Some dipterous galls of Southern California. *Journal of Ento. and Zool, Pomona* 17(1):1-8.

Brown, Leland R., and Eads, Clark O. 1965. *A technical study of insects affecting the oak tree in Southern California.* California Agricultural Experiment Station, Univ. of Calif., Bulletin 810, pp. 79-91.

Burnett, John A. 1974. A new cynipid wasp from California. *Pan Pacific Ento.* 50(3):298-302.

Caltagirone, L. E. 1964. Notes on the biology, parasites, and inquilines of *Pontania pacifica* (Hymenoptera:Tenthredinidae) a leaf gall incitant on *Salix lasiolepis. Annals of the Ento. Soc. of Amer.* 57(3):279-291.

Cockerell, T. D. A. 1905. An aphid gall on *Arctostaphylos. Canadian Ento.* 37(11):391-392.

Dailey, D. Charles. 1969. Synonymy of *Dryocosmus attractans* and *Callirhytis uvellae. Pan Pacific Ento.* 45(2):132-134.

———., and Sprenger, C. M. 1973a. Unisexual generation of *Andricus atrimentus* (Hymenoptera:Cynipidae). *Pan Pacific Ento.* 49(2):171-173.

———., and Campbell, L. 1973b. A new species of *Diplolepis* from California (Hymenoptera:Cynipidae). *Pan Pacific Ento.* 49(2):174-176.

———., and Sprenger, C. M. 1973c. Synonymy of *Andricus gigas* and the bisexual generation of *Andricus crenatus* (Hymenoptera:Cynipidae). *Pan Pacific Ento.* 49(3):188-191.

———., Perry, T., and Sprenger, C. M. 1974. Biology of three *Callirhytis* gall wasps from Pacific slope Erythrobalanus oaks. *Pan Pacific Ento.* 50(1): 60-67.

Darlington, Arnold. 1968. *The pocket encyclopedia of plant galls.* London: Blandford Press.

Doutt, R. L. 1948. The suitability of insect conditioned plant tissues as habitats for successive insect species. *Pan Pacific Ento.* 24(3):121-122.

———. 1960. Heterogeny in *Andricus crystallinus* (Hymenoptera:Cynipidae). *Pan Pacific Ento.* 36(4):167-170.

Essig, E. O. 1926. *Insects of western North America.* New York:Macmillan.

Felt, E. P. 1916. New western gall midges. *Journal of New York Ento. Soc.* 24(3):175-196.

———. 1965. *Plant galls and gall makers.* New York: Hafner.

Ferris, Roxanna S. 1962. *Death Valley wildflowers.* Death Valley: Death Valley Natural History Association.

Force, Don C. 1974. Ecology of insect host-parasitoid communities. *Science* 184:624-632.

Fronk, W. D., Beetle, A. A., and Fullerton, D. G. 1964. Dipterous galls on the *Artemesia tridentata* complex and insects associated with them. *Annals of the Ento. Soc. of Amer.* 57(5):575-577.

Gagne, Raymond J. 1969. A review of the genus *Walshomyia* including a new species reared from *Cupressus* galls in California. *Pan Pacific Ento.* 45(1): 16-19.

———. 1973. A generic synopsis of nearctic Cecidomyiidi (Diptera: Cecidomyiidae). *Annals of Ento. Soc. of Amer.* 66(4):857-889.

———. 1975. A redefinition of *Diarthronomyia* felt as a subgenus of *Rhopalomyia Rubsaamen* (Diptera:Cecidomyiidae). *Annals of the Ento. Soc. of Amer.* 68(3):482-484.

Grigarick, A. A., and Lange, W. H. 1968. Seasonal development and emergence of two species of gall-forming aphids, *Pemphigus bursarius* and *P. nortoni,* associated with poplar trees in California. *Annals of Ento. Soc. of Amer.* 61(2):509-514.

Harper, A. M. 1959. Gall aphids on poplar in Alberta, descriptions of galls and distributions of aphids. *Canadian Ento.* 91(8):489-496.

———. 1966. Three additional poplar gall aphids from southern Alberta. *Canadian Ento.* 98(11):1212-1214.

Harvey, John M. 1948. Ecology of an itonidid fly associated with a rust on *Baccharis pilularis consanguinea. Pan Pacific Ento.* 24(4):194-200.

Hepting, George H. 1971. *Diseases of forest and shade trees of the U.S.* U.S. Dept. of Agriculture Handbook 386.

Hildebrand, D. C., and Schroth, M. N. 1967. A new species of Erwinia causing the drippy nut disease of live oaks. *Phytopathology* 57(3):250-253.

Hovanitz, William. 1958. Insects and plant galls. *Scientific American* (November):151-162.

Hutchins, Ross E. 1969. *Galls and gall insects.* New York:Dodd, Mead & Co.

Jepson, L. R. 1975. *Mites injurious to economic plants.* Berkeley: Univ. of Calif. Press.

Johnson, Warren T., and Lyon, Howard H. 1976. *Insects that feed on trees and shrubs.* Ithaca, N.Y.:Cornell Univ. Press:Comstock Publishing Co.

Keifer, H. H. 1952. The eriophyid mites of California. *Univ. of Calif. Press, Bull. of the Calif. Insect Survey* 2(1).

Kinsey, Alfred. 1922. Studies of some new and described Cynipidae. *Indiana Univ. Studies* 9(53):1-141.

Lange, W. H. 1965. *Biosystematics of American Pemphigus (Homoptera: Aphidoidea).* 12th International Congress of Entomology, pp. 102-104.

Lyon, Robert J. 1959. An alternating sexual generation in the gall wasp *Callirhytis pomiformis. Bull. Southern Calif. Acad. of Sci.* 58(1):33-37.

———. 1963. An alternate generation of *Heteroecus pacificus. Proc. of the Ento. Soc. of Wash.* 65(3):250-254.

———. 1964. An alternate generation of *Callirhytis agrifoliae. Proc. of the Ento. Soc. of Wash.* 66(3):193-196.

———. 1969. The alternate generation of *Callirhytis quercussuttoni. Proc. of the Ento. Soc. of Wash.* 71(1):61-65.

———. 1970. Heterogeny in *Callirhytis serricornis. Proc. of the Ento. Soc. of Wash.* 72(2):176-178.

Mani, M. S. 1964. *Ecology of plant galls.* The Hague:Dr. W. Junk.

McMinn, Howard E., and Maino, Evelyn. 1963. *Pacific coast trees.* Berkeley: Univ. of Calif. Press.

Mix, A. J. 1949. Monograph of the genus *Taphrina. Univ. of Kansas Science Bull.* 33(1):3-167.

Munz, Phillip A. 1963. *A flora of California.* Berkeley:Univ. of Calif. Press.

———. 1968. *Supplement to a California flora.* Berkeley:Univ. of Calif. Press.

Oldfield, G. N., Hobsa, R. F., and Wilson, N. S. 1970. Discovery and characterization of spermatophores in the Eriophyoidea. *Annals of Ento. Soc. of Amer.* 63(2):520-526.

O'Reilly, Henry J., Ross, Norman W., and Post, George R. 1968. *Leaf curl in peaches and nectarines.* Univ. of Calif. Agricultural Extention Service, Publication AXT-263.

Palmer, Miriam A. 1952. *Aphids of the Rocky Mountains region.* Thomas Say Foundation Publisher.

Paulus, Albert D., and McCain, Arthur H. 1965. *Control of bacteria gall of oleander.* Univ. of Calif. Agricultural Extension Service, Publication AXT-129.

Pritchard, A. Earl. 1953. The gall midges of California. *Univ. of Calif. Press, Bull. of Calif. Insect Survey* 2(2).

Rosenthal, Sara S. 1968. "Biology and host relations of some Cynipidae forming galls on Quercus." Ph.D. dissertation, Univ. of Calif, Berkeley.

———., and Koehler, C. S. 1971a. Heterogeny in some gall forming Cynipidae (Hymenoptera) with notes on the biology of *Neuroterus saltatorius. Annals of the Ento. Soc. of Amer.* 64(3):565-570.

———., and Koehler, C. S. 1971b. Intertree distributions of some cynipid (Hymenoptera) galls on *Quercus lobata. Annals of the Ento. Soc. of Amer.* 64(3):571-574.

Ross, N., Schroth, M. N., Sanborn, R., O'Reilly, H. J., and Thompson, J. P. 1970. Reducing loss from crown gall disease. *California Agricultural Experiment Station, Univ. of Calif. Bull.* 845:3-10.

Shorthouse, J. D. 1973a. *Common insect galls of Saskatchewan.* Blue Jay.

———. 1973b. The insect community associated with rose galls of *Diplolepis polita* (Hymenoptera:Cynipidae). *Quaestiones entomologicas* 9:55-98.

Smith, Edward L. 1968. Biosystematics and morphology of Symphyta. I. Stem-galling Euura of the California region and a new female genitalic nomenclature. *Annals of the Ento. Soc. of Amer.* 61(6):1389-1407.

———. 1970. Biosystematics and morphology of Symphyta. II. Biology of gall making nematine sawflies in the California region. *Annals of the Ento. Soc. of Amer.* 63(1):36-51.

Tilden, James W. 1951. *The insect associates of Baccharis pilularis.* Microentomology, vol. 16, part 1, pp. 149-188. Stanford: Stanford Univ. Press.

Townsend, C. H. Tyler. 1893. Fleshy cecidomyiid twig gall on *Atriplex canescens. American Naturalist* 27:1021.

Wangberg, James K. 1975. Biology of the thimbleberry gallmaker *Diastrophus kincaidii* (Hymenoptera:Cynipidae). *Pan Pacific Ento.* 51(1):39-48.

Weld, Lewis H. 1926. Field notes on gall-inhabiting cynipid wasps with descriptions of new species. *Proc. of U.S. National Museum* 68(10).

———. 1952. New American cynipid wasps from galls. *Proc. of U.S. National Museum* 102:316-342.

———. 1957. *Cynipid galls of the Pacific slope.* Ann Arbor, Michigan.

Yarwood, C. E. 1968. Tillage and plant diseases. *Bioscience* 18(1):27-30.

———. 1969. Tillage increases plant diseases. *California Agriculture* (March): 4-6.

A

B

C